FREE Test Taking Tips DVD Offer

To help us better serve you, we have developed a Test Taking Tips DVD that we would like to give you for FREE. **This DVD covers world-class test taking tips that you can use to be even more successful when you are taking your test.**

All that we ask is that you email us your feedback about your study guide. Please let us know what you thought about it – whether that is good, bad or indifferent.

To get your **FREE Test Taking Tips DVD**, email freedvd@studyguideteam.com with “FREE DVD” in the subject line and the following information in the body of the email:

a. The title of your study guide.

b. Your product rating on a scale of 1-5, with 5 being the highest rating.

c. Your feedback about the study guide. What did you think of it?

d. Your full name and shipping address to send your free DVD.

If you have any questions or concerns, please don’t hesitate to contact us at freedvd@studyguideteam.com.

Thanks again!

CBEST Prep Practice Book 2020 & 2021

3 CBEST Practice Tests [2nd Edition]

Test Prep Books

Table of Contents

Quick Overview

As you draw closer to taking your exam, effective preparation becomes more and more important. Thankfully, you have this study guide to help you get ready. Use this guide to help keep your studying on track and refer to it often.

This study guide contains several key sections that will help you be successful on your exam. The guide contains tips for what you should do the night before and the day of the test. Also included are test-taking tips. Knowing the right information is not always enough. Many well-prepared test takers struggle with exams. These tips will help equip you to accurately read, assess, and answer test questions.

A large part of the guide is devoted to showing you what content to expect on the exam and to helping you better understand that content. In this guide are practice test questions so that you can see how well you have grasped the content. Then, answer explanations are provided so that you can understand why you missed certain questions.

Don't try to cram the night before you take your exam. This is not a wise strategy for a few reasons. First, your retention of the information will be low. Your time would be better used by reviewing information you already know rather than trying to learn a lot of new information. Second, you will likely become stressed as you try to gain a large amount of knowledge in a short amount of time. Third, you will be depriving yourself of sleep. So be sure to go to bed at a reasonable time the night before. Being well-rested helps you focus and remain calm.

Be sure to eat a substantial breakfast the morning of the exam. If you are taking the exam in the afternoon, be sure to have a good lunch as well. Being hungry is distracting and can make it difficult to focus. You have hopefully spent lots of time preparing for the exam. Don't let an empty stomach get in the way of success!

When travelling to the testing center, leave earlier than needed. That way, you have a buffer in case you experience any delays. This will help you remain calm and will keep you from missing your appointment time at the testing center.

Be sure to pace yourself during the exam. Don't try to rush through the exam. There is no need to risk performing poorly on the exam just so you can leave the testing center early. Allow yourself to use all of the allotted time if needed.

Remain positive while taking the exam even if you feel like you are performing poorly. Thinking about the content you should have mastered will not help you perform better on the exam.

Once the exam is complete, take some time to relax. Even if you feel that you need to take the exam again, you will be well served by some down time before you begin studying again. It's often easier to convince yourself to study if you know that it will come with a reward!

Test-Taking Strategies

1. Predicting the Answer

When you feel confident in your preparation for a multiple-choice test, try predicting the answer before reading the answer choices. This is especially useful on questions that test objective factual knowledge. By predicting the answer before reading the available choices, you eliminate the possibility that you will be distracted or led astray by an incorrect answer choice. You will feel more confident in your selection if you read the question, predict the answer, and then find your prediction among the answer choices. After using this strategy, be sure to still read all of the answer choices carefully and completely. If you feel unprepared, you should not attempt to predict the answers. This would be a waste of time and an opportunity for your mind to wander in the wrong direction.

2. Reading the Whole Question

Too often, test takers scan a multiple-choice question, recognize a few familiar words, and immediately jump to the answer choices. Test authors are aware of this common impatience, and they will sometimes prey upon it. For instance, a test author might subtly turn the question into a negative, or he or she might redirect the focus of the question right at the end. The only way to avoid falling into these traps is to read the entirety of the question carefully before reading the answer choices.

3. Looking for Wrong Answers

Long and complicated multiple-choice questions can be intimidating. One way to simplify a difficult multiple-choice question is to eliminate all of the answer choices that are clearly wrong. In most sets of answers, there will be at least one selection that can be dismissed right away. If the test is administered on paper, the test taker could draw a line through it to indicate that it may be ignored; otherwise, the test taker will have to perform this operation mentally or on scratch paper. In either case, once the obviously incorrect answers have been eliminated, the remaining choices may be considered. Sometimes identifying the clearly wrong answers will give the test taker some information about the correct answer. For instance, if one of the remaining answer choices is a direct opposite of one of the eliminated answer choices, it may well be the correct answer. The opposite of obviously wrong is obviously right! Of course, this is not always the case. Some answers are obviously incorrect simply because they are irrelevant to the question being asked. Still, identifying and eliminating some incorrect answer choices is a good way to simplify a multiple-choice question.

4. Don't Overanalyze

Anxious test takers often overanalyze questions. When you are nervous, your brain will often run wild, causing you to make associations and discover clues that don't actually exist. If you feel that this may be a problem for you, do whatever you can to slow down during the test. Try taking a deep breath or counting to ten. As you read and consider the question, restrict yourself to the particular words used by the author. Avoid thought tangents about what the author *really* meant, or what he or she was *trying* to say. The only things that matter on a multiple-choice test are the words that are actually in the question. You must avoid reading too much into a multiple-choice question, or supposing that the writer meant something other than what he or she wrote.

5. No Need for Panic

It is wise to learn as many strategies as possible before taking a multiple-choice test, but it is likely that you will come across a few questions for which you simply don't know the answer. In this situation, avoid panicking. Because most multiple-choice tests include dozens of questions, the relative value of a single wrong answer is small. As much as possible, you should compartmentalize each question on a multiple-choice test. In other words, you should not allow your feelings about one question to affect your success on the others. When you find a question that you either don't understand or don't know how to answer, just take a deep breath and do your best. Read the entire question slowly and carefully. Try rephrasing the question a couple of different ways. Then, read all of the answer choices carefully. After eliminating obviously wrong answers, make a selection and move on to the next question.

6. Confusing Answer Choices

When working on a difficult multiple-choice question, there may be a tendency to focus on the answer choices that are the easiest to understand. Many people, whether consciously or not, gravitate to the answer choices that require the least concentration, knowledge, and memory. This is a mistake. When you come across an answer choice that is confusing, you should give it extra attention. A question might be confusing because you do not know the subject matter to which it refers. If this is the case, don't eliminate the answer before you have affirmatively settled on another. When you come across an answer choice of this type, set it aside as you look at the remaining choices. If you can confidently assert that one of the other choices is correct, you can leave the confusing answer aside. Otherwise, you will need to take a moment to try to better understand the confusing answer choice. Rephrasing is one way to tease out the sense of a confusing answer choice.

7. Your First Instinct

Many people struggle with multiple-choice tests because they overthink the questions. If you have studied sufficiently for the test, you should be prepared to trust your first instinct once you have carefully and completely read the question and all of the answer choices. There is a great deal of research suggesting that the mind can come to the correct conclusion very quickly once it has obtained all of the relevant information. At times, it may seem to you as if your intuition is working faster even than your reasoning mind. This may in fact be true. The knowledge you obtain while studying may be retrieved from your subconscious before you have a chance to work out the associations that support it. Verify your instinct by working out the reasons that it should be trusted.

8. Key Words

Many test takers struggle with multiple-choice questions because they have poor reading comprehension skills. Quickly reading and understanding a multiple-choice question requires a mixture of skill and experience. To help with this, try jotting down a few key words and phrases on a piece of scrap paper. Doing this concentrates the process of reading and forces the mind to weigh the relative importance of the question's parts. In selecting words and phrases to write down, the test taker thinks about the question more deeply and carefully. This is especially true for multiple-choice questions that are preceded by a long prompt.

9. Subtle Negatives

One of the oldest tricks in the multiple-choice test writer's book is to subtly reverse the meaning of a question with a word like *not* or *except*. If you are not paying attention to each word in the question, you can easily be led astray by this trick. For instance, a common question format is, "Which of the following is...?" Obviously, if the question instead is, "Which of the following is not...?," then the answer will be quite different. Even worse, the test makers are aware of the potential for this mistake and will include one answer choice that would be correct if the question were not negated or reversed. A test taker who misses the reversal will find what he or she believes to be a correct answer and will be so confident that he or she will fail to reread the question and discover the original error. The only way to avoid this is to practice a wide variety of multiple-choice questions and to pay close attention to each and every word.

10. Reading Every Answer Choice

It may seem obvious, but you should always read every one of the answer choices! Too many test takers fall into the habit of scanning the question and assuming that they understand the question because they recognize a few key words. From there, they pick the first answer choice that answers the question they believe they have read. Test takers who read all of the answer choices might discover that one of the latter answer choices is actually *more* correct. Moreover, reading all of the answer choices can remind you of facts related to the question that can help you arrive at the correct answer. Sometimes, a misstatement or incorrect detail in one of the latter answer choices will trigger your memory of the subject and will enable you to find the right answer. Failing to read all of the answer choices is like not reading all of the items on a restaurant menu: you might miss out on the perfect choice.

11. Spot the Hedges

One of the keys to success on multiple-choice tests is paying close attention to every word. This is never truer than with words like almost, most, some, and sometimes. These words are called "hedges" because they indicate that a statement is not totally true or not true in every place and time. An absolute statement will contain no hedges, but in many subjects, the answers are not always straightforward or absolute. There are always exceptions to the rules in these subjects. For this reason, you should favor those multiple-choice questions that contain hedging language. The presence of qualifying words indicates that the author is taking special care with his or her words, which is certainly important when composing the right answer. After all, there are many ways to be wrong, but there is only one way to be right! For this reason, it is wise to avoid answers that are absolute when taking a multiple-choice test. An absolute answer is one that says things are either all one way or all another. They often include words like *every*, *always*, *best*, and *never*. If you are taking a multiple-choice test in a subject that doesn't lend itself to absolute answers, be on your guard if you see any of these words.

12. Long Answers

In many subject areas, the answers are not simple. As already mentioned, the right answer often requires hedges. Another common feature of the answers to a complex or subjective question are qualifying clauses, which are groups of words that subtly modify the meaning of the sentence. If the question or answer choice describes a rule to which there are exceptions or the subject matter is complicated, ambiguous, or confusing, the correct answer will require many words in order to be expressed clearly and accurately. In essence, you should not be deterred by answer choices that seem excessively long. Oftentimes, the author of the text will not be able to write the correct answer without offering some qualifications and modifications. Your job is to read the answer choices thoroughly and completely and to select the one that most accurately and precisely answers the question.

13. Restating to Understand

Sometimes, a question on a multiple-choice test is difficult not because of what it asks but because of how it is written. If this is the case, restate the question or answer choice in different words. This process serves a couple of important purposes. First, it forces you to concentrate on the core of the question. In order to rephrase the question accurately, you have to understand it well. Rephrasing the question will concentrate your mind on the key words and ideas. Second, it will present the information to your mind in a fresh way. This process may trigger your memory and render some useful scrap of information picked up while studying.

14. True Statements

Sometimes an answer choice will be true in itself, but it does not answer the question. This is one of the main reasons why it is essential to read the question carefully and completely before proceeding to the answer choices. Too often, test takers skip ahead to the answer choices and look for true statements. Having found one of these, they are content to select it without reference to the question above. Obviously, this provides an easy way for test makers to play tricks. The savvy test taker will always read the entire question before turning to the answer choices. Then, having settled on a correct answer choice, he or she will refer to the original question and ensure that the selected answer is relevant. The mistake of choosing a correct-but-irrelevant answer choice is especially common on questions related to specific pieces of objective knowledge. A prepared test taker will have a wealth of factual knowledge at his or her disposal, and should not be careless in its application.

15. No Patterns

One of the more dangerous ideas that circulates about multiple-choice tests is that the correct answers tend to fall into patterns. These erroneous ideas range from a belief that B and C are the most common right answers, to the idea that an unprepared test-taker should answer "A-B-A-C-A-D-A-B-A." It cannot be emphasized enough that pattern-seeking of this type is exactly the WRONG way to approach a multiple-choice test. To begin with, it is highly unlikely that the test maker will plot the correct answers according to some predetermined pattern. The questions are scrambled and delivered in a random order. Furthermore, even if the test maker was following a pattern in the assignation of correct answers, there is no reason why the test taker would know which pattern he or she was using. Any attempt to discern a pattern in the answer choices is a waste of time and a distraction from the real work of taking the test. A test taker would be much better served by extra preparation before the test than by reliance on a pattern in the answers.

FREE DVD OFFER

Don't forget that doing well on your exam includes both understanding the test content and understanding how to use what you know to do well on the test. We offer a completely FREE Test Taking Tips DVD that covers world class test taking tips that you can use to be even more successful when you are taking your test.

All that we ask is that you email us your feedback about your study guide. To get your **FREE Test Taking Tips DVD**, email freedvd@studyguideteam.com with "FREE DVD" in the subject line and the following information in the body of the email:

- The title of your study guide.
- Your product rating on a scale of 1-5, with 5 being the highest rating.
- Your feedback about the study guide. What did you think of it?
- Your full name and shipping address to send your free DVD.

Introduction to the CBEST

Function of the Test

The California Basic Educational Skills Test (CBEST) was created by the government of the state of California as a way for teacher candidates to demonstrate proficiency in reading, writing, and mathematics. Individuals applying for their first California teaching credential, applying for admission to certain teacher preparation and credentialing programs, or seeking employment in a California school district or educational agency must meet the California Basic Skills Requirement. The California legislature has established eight different ways to meet this requirement, such as achieving certain scores on various tests, one of which is to pass the CBEST. The state of Nevada has also adopted the CBEST as a way to meet certain Nevada licensing requirements.

CBEST scores are generally used only for California and Nevada teacher licensing, credentialing and hiring. In the 2014-2015 school year, 32,890 individuals took the CBEST for the first time. 22,847 of these first-time test takers passed, for a passing rate of 69.5%. The passing rate over the last five years has ranged from 69.5% to 71.4%. Of the three sections, reading and math are generally a bit easier to pass, with passing rates of around 80%, while writing is a bit more difficult, with a passing rate around 73%. (http://www.ctc.ca.gov/commission/agendas/2016-04/2016-04-5C.pdf).

Test Administration

The computer-based version of the CBEST is offered year-round by appointment, Monday through Saturday, excluding certain holidays, at Pearson VUE testing centers. The paper-based version is available on a more limited basis, usually around five times per school year. Individuals who do not pass must wait 45 days to attempt the CBEST again on computer, or may attempt the test again on any scheduled paper-administered test day.

All CBEST test sites are wheelchair-accessible. Individuals with documented disabilities may receive additional accommodations such as allowance of a medical device in the testing room, additional breaks, and use of tools such as a magnifying glass or straight edge. Individuals seeking such accommodation should complete and submit an Alternative Testing Arrangements Request Form along with appropriate documentation prior to their scheduled test date. The form must be submitted each time a student seeking accommodations takes the CBEST exam.

Test Format

A CBEST testing session will last four hours. Test takers may elect to take one, two, or all three of the sections offered. The reading section is made up of questions designed to assess the test takers ability to comprehend information in written passages, tables, and graphs. The mathematics section is primarily comprised of word problems to be solved without the use of a calculator. The writing section is composed of two essays, one in which the test taker is asked to analyze a given situation or statement and one in which the test taker is asked to describe a personal experience

A summary of the content of the CBEST is as follows:

Section	Subsection	Approx # of Questions
Reading	Critical analysis and evaluation	20
	Comprehension and research skills	30
Mathematics	Estimation, measurement, and statistical principles	15
	Computation and problem solving	17
	Numerical and graphic relationships	17
Writing	Essay	2

Scoring

Scoring on the reading and math sections is done by calculating a raw score based on the number of correct answers with no penalty for guessing incorrectly and converting that raw score to a scaled score between 20 and 80. Similarly, the essays in the writing section are given a score between 1 and 4 by two readers for each of the two essays. The total raw score (between 4 and 16) is then converted to a scaled score between 20 and 80.

Test takers may pass the CBEST by *either* earning scaled score on each section of at least 41, *or* earning a scaled score on each section of at least 37 and a total scaled score of at least 123. As discussed above, the passing rate on individual sections ranges from the mid-70s to around 80%, and the overall passing rate hovers around 70%.

CBEST Practice Test #1

Reading

Questions 1–3 refer to the following paragraph.

> The Brookside area is an older part of Kansas City, developed mainly in the 1920s and 30s, and is considered one of the nation's first "planned" communities with shops, restaurants, parks, and churches all within a quick walk. A stroll down any street reveals charming two-story Tudor and Colonial homes with smaller bungalows sprinkled throughout the beautiful tree-lined streets. It is common to see lemonade stands on the corners and baseball games in the numerous "pocket" parks tucked neatly behind rows of well-manicured houses. The Brookside shops on 63rd street between Wornall Road and Oak Street are a hub of commerce and entertainment where residents freely shop and dine with their pets (and children) in tow. This is also a common "hangout" spot for younger teenagers because it is easily accessible by bike for most. In short, it is an idyllic neighborhood just minutes from downtown Kansas City.

1. Which of the following states the main idea of this paragraph?
    a. The Brookside shops are a popular hangout for teenagers.
    b. There are a number of pocket parks in the Brookside neighborhood.
    c. Brookside is a great place to live.
    d. Brookside has a high crime rate.
    e. Everyone should move to Brookside.

2. In what kind of publication might you read the above paragraph?
    a. Fictional novel
    b. Literary journal
    c. Newspaper article
    d. Movie review
    e. Community profile

3. According to this paragraph, which of the following is unique to this neighborhood?
    a. It is old.
    b. It is in Kansas City.
    c. It has shopping.
    d. It is one of the nation's first planned communities.
    e. It has both Tudor and Colonial homes.

Questions 4–6 refer to the following excerpt from a government publication table of contents.

Contents

From http://purl.fdlp.gov/GPO/gpo66588

4. In which chapter would you find information for a research paper about how the nation plans to address the problem of child neglect in affected communities?
   a. Chapter 3
   b. Chapter 4
   c. Chapter 5
   d. Chapter 6
   e. Chapter 7

5. On which page would you find information about the needs of the American Indian child?
   a. 52
   b. 60
   c. 86
   d. 106
   e. 120

6. According to the table of contents, Chapter 1 is titled "Confronting the Tragedy of Child Abuse and Neglect Fatalities." Based on the chapter title and sections, which of the following is a good summary of this chapter?
   a. Children's lives today are valuable and worth saving, and the children of the future are just as important.
   b. The outlining of a plan to solve the problem of child neglect in disproportionately affected communities is one of the main goals here.
   c. Child abuse and neglect is alive and rampant in our communities, and this chapter outlines those who are in dire need of our attention.
   d. This chapter outlines resources for families who are affected by child abuse and neglect.
   e. Provides quantitative facts in order to help make sense of the issue of child abuse.

Questions 7–9 refer to the following excerpt taken from Walt Whitman's "Letters from a Traveling Bachelor," published in the New York Sunday Dispatch *on 14 October 1849.*

> At its easternmost part, Long Island opens like the upper and under jaws of some prodigious alligator; the upper and larger one terminating in Montauk Point. The bay that lies in here, and part of which forms the splendid harbor of Greenport, where the Long Island Railroad ends, is called Peconic Bay; and a beautiful and varied water is it, fertile in fish and feathered game. I, who am by no means a skillful fisherman, go down for an hour of a morning on one of the docks, or almost anywhere along shore, and catch a mess of black-fish, which you couldn't buy in New York for a dollar—large fat fellows, with meat on their bones that it takes a pretty long fork to stick through. They have a way here of splitting these fat black-fish and poggies, and broiling them on the coals, beef-steak-fashion, which I recommend your Broadway cooks to copy.

7. Whitman's comparison of the easternmost part of Long Island to an alligator is an example of which literary device?
   a. Hyperbole
   b. Metaphor
   c. Personification
   d. Alliteration
   e. Simile

8. Which of the following is the best summary of this passage?
   a. Walt Whitman was impressed with the quantity and quality of fish he found in Peconic Bay.
   b. Walt Whitman prefers the fish found in restaurants in New York.
   c. Walt Whitman was a Broadway chef.
   d. Walt Whitman is frustrated because he is not a very skilled fisherman.
   e. Walt Whitman is describing the geography of Long Island.

9. Using the context clues in the passage, what is the meaning of the word *prodigious* in the first line?
   a. Out of place
   b. Great in extent, size, or degree
   c. Mean-spirited
   d. Extremely intelligent
   e. Of a somber disposition

10. Which of the following would be a good topic sentence if you were writing a paragraph about the effects of education on crime rates?
   a. Crime statistics are difficult to verify for a number of reasons.
   b. Educated people do not commit crimes.
   c. Education is an important factor in lowering crime rates.
   d. Education has been proven to lower crime rates by as much as 20% in some urban areas.
   e. It is untrue that areas that have higher education are higher in crime rates.

11. Which of the following is not an example of a good thesis statement for an essay?
   a. Animals in danger of becoming extinct come from a wide range of countries.
   b. Effective leadership requires specific qualities that anyone can develop.
   c. Cashew milk is perfect for bakers who want to use dairy-free alternatives.
   d. In order to fully explore the wreck of the Titanic, scientists must address several problems.
   e. Industrial waste poured into Lake Michigan has killed 27 percent of marine life in the past decade.

Questions 12–15 refer to the following passage, titled "Education is Essential to Civilization."

Early in my career, a master teacher shared this thought with me: "Education is the last bastion of civility." While I did not completely understand the scope of those words at the time, I have since come to realize the depth, breadth, truth, and significance of what he said. Education provides society with a vehicle for raising its children to be civil, decent, human beings with something valuable to contribute to the world. It is really what makes us human and what distinguishes us as civilized creatures.

Being "civilized" humans means being "whole" humans. Education must address the mind, body, and soul of students. It would be detrimental to society if our schools were myopic in their focus, only meeting the needs of the mind. As humans, we are multi-dimensional, multi-faceted beings who need more than head knowledge to survive. The human heart and psyche have to be fed in order for the mind to develop properly, and the body must be maintained and exercised to help fuel the working of the brain.

Education is a basic human right, and it allows us to sustain a democratic society in which participation is fundamental to its success. It should inspire students to seek better solutions to world problems and to dream of a more equitable society. Education should never discriminate on any basis, and it should create individuals who are self-sufficient, patriotic, and tolerant of other's ideas.

All children can learn, although not all children learn in the same manner. All children learn best, however, when their basic physical needs are met, and they feel safe, secure, and loved. Students are much more responsive to a teacher who values them and shows them respect as individual people. Teachers must model at all times the way they expect students to treat them and their peers. If teachers set high expectations for their students, the students will rise to that high level. Teachers must make the well-being of their students their primary focus and must not be afraid to let their students learn from their own mistakes.

In the modern age of technology, a teacher's focus is no longer the "what" of the content, but more importantly, the "why." Students are bombarded with information and have access to ANY information they need right at their fingertips. Teachers have to work harder than ever before to help students identify salient information and to think critically about the information they encounter. Students have to read between the lines, identify bias, and determine who they can trust in the milieu of ads, data, and texts presented to them.

Schools must work in consort with families in this important mission. While children spend most of their time in school, they are dramatically and indelibly shaped by the influences of their family and culture. Teachers must not only respect this fact but must strive to include parents in the education of their children and must work to keep parents informed of progress and problems. Communication between classroom and home is essential for a child's success.

Humans have always aspired to be more, do more, and to better ourselves and our communities. This is where education lies, right at the heart of humanity's desire to be all that we can be. Education helps us strive for higher goals and better treatment of ourselves and others. I shudder to think what would become of us if education ceased to be the "last bastion of civility." We must be unapologetic about expecting excellence from our students—our very existence depends upon it.

12. Which of the following best summarizes the author's main point?
   a. Education as we know it is over-valued in modern society, and we should find alternative solutions.
   b. The survival of the human race depends on the educational system, and it is worth fighting for to make it better.
   c. The government should do away with all public schools and require parents to home school their children instead.
   d. While education is important, some children simply are not capable of succeeding in a traditional classroom.
   e. Students are learning new ways to learn while teachers are learning to adapt to their needs.

13. Based on this passage, which of the following can be inferred about the author?
   a. The author feels passionately about education.
   b. The author does not feel strongly about their point.
   c. The author is angry at the educational system.
   d. The author is unsure about the importance of education.
   e. The author does not trust the government.

14. Based on this passage, which of the following conclusions could be drawn about the author?
   a. The author would not support raising taxes to help fund much needed reforms in education.
   b. The author would support raising taxes to help fund much needed reforms in education, as long as those reforms were implemented in higher socio-economic areas first.
   c. The author would support raising taxes to help fund much needed reforms in education for all children in all schools.
   d. The author would support raising taxes only in certain states to help fund much needed reforms in education.
   e. The author would not support raising taxes unless the people in the communities agreed to it.

15. According to the passage, which of the following is not mentioned as an important factor in education today?
   a. Parent involvement
   b. Communication between parents and teachers
   c. Impact of technology
   d. Cost of textbooks
   e. Safe and healthy children

For question 16, use the following graphics.

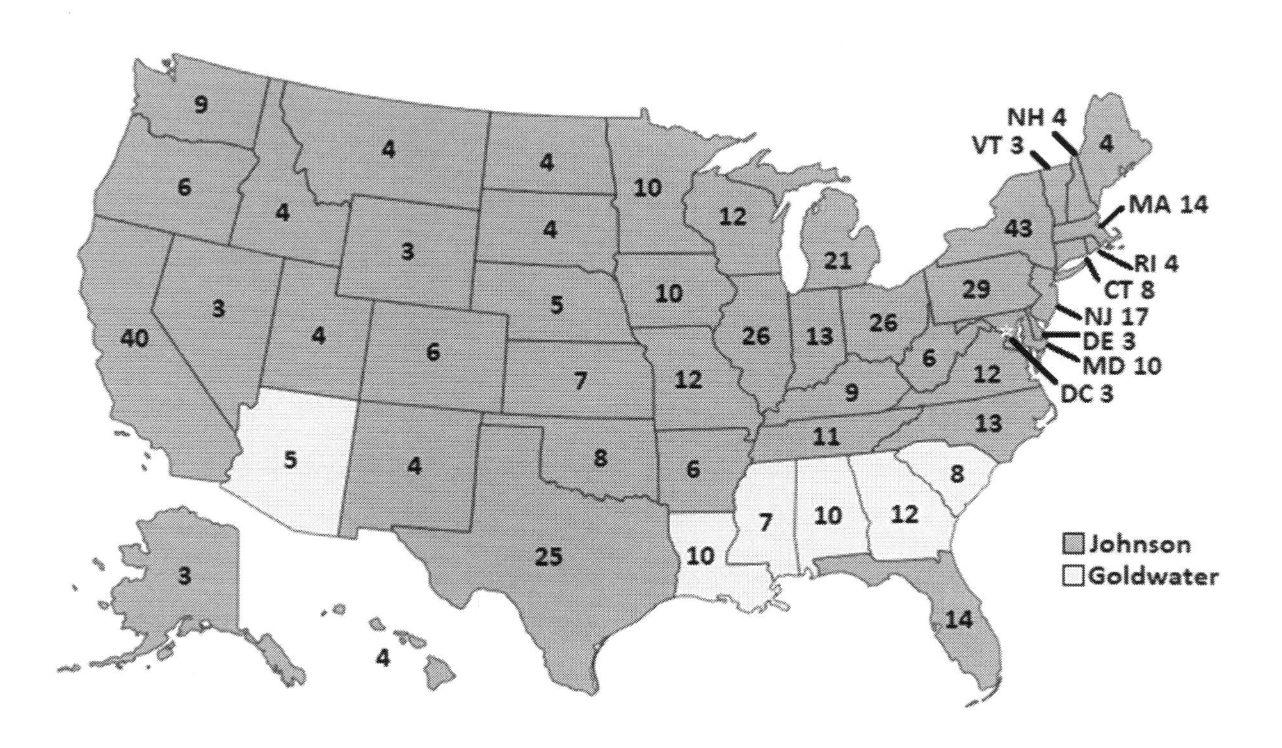

1964 Election Results			
Candidate	**Party**	**Electoral Votes**	**Popular Votes**
Lyndon B. Johnson	Democratic	486	42,825,463
Barry M. Goldwater	Republican	52	27,146,969

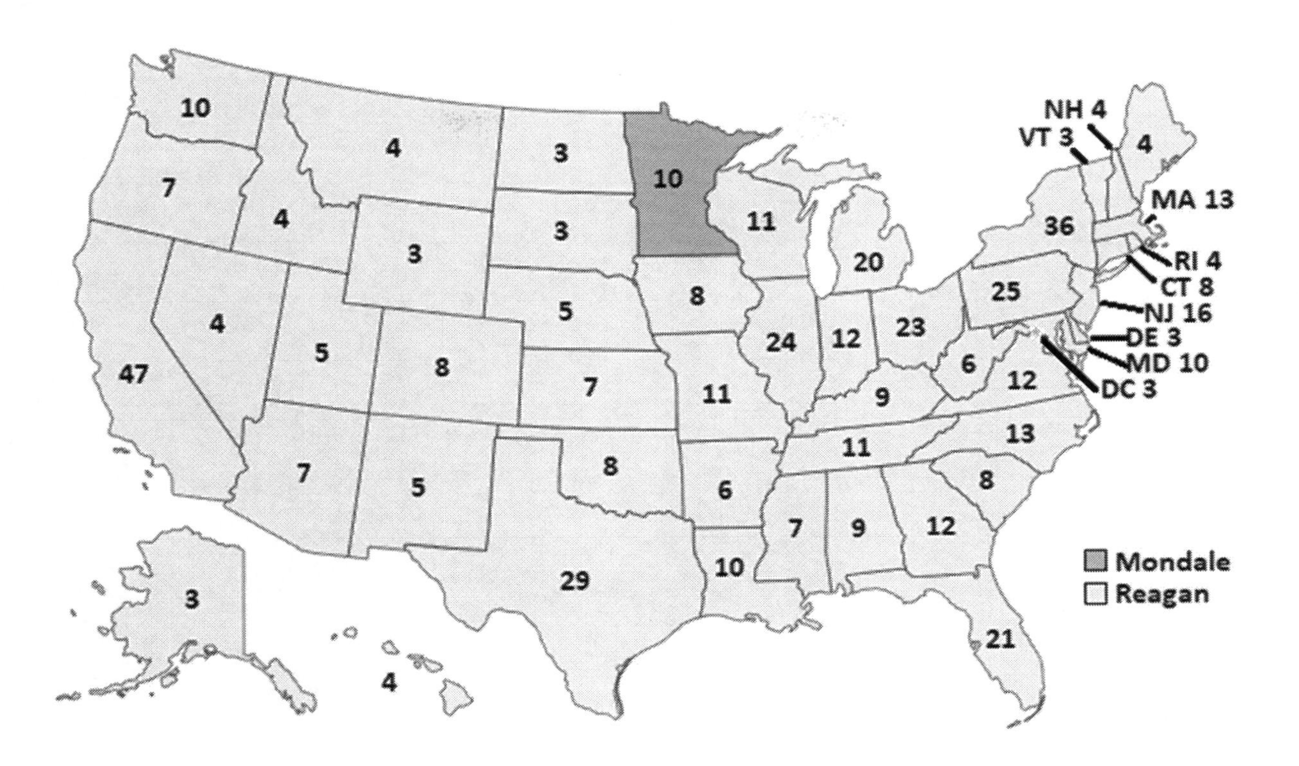

1984 Election Results			
Candidate	**Party**	**Electoral Votes**	**Popular Votes**
Ronald Reagan (I)	Republican	525	54,455,000
Walter F. Mondale	Democratic	13	37,577,000

16. Based on the two maps, which of the following statements is true?
   a. Over twenty years, the country's voter turnout for a presidential election stayed the exact same.
   b. Over twenty years, there was a noticeable decrease in voter turnout during a presidential election.
   c. Over twenty years, the electoral vote swung from almost completely Democrat to almost completely Republican.
   d. There was very little change in the number of electoral votes won for each party from 1964 to 1984.
   e. Over twenty years, the electoral vote swung from almost completely Republican to almost completely Democrat.

For question 17, use the following graphic.

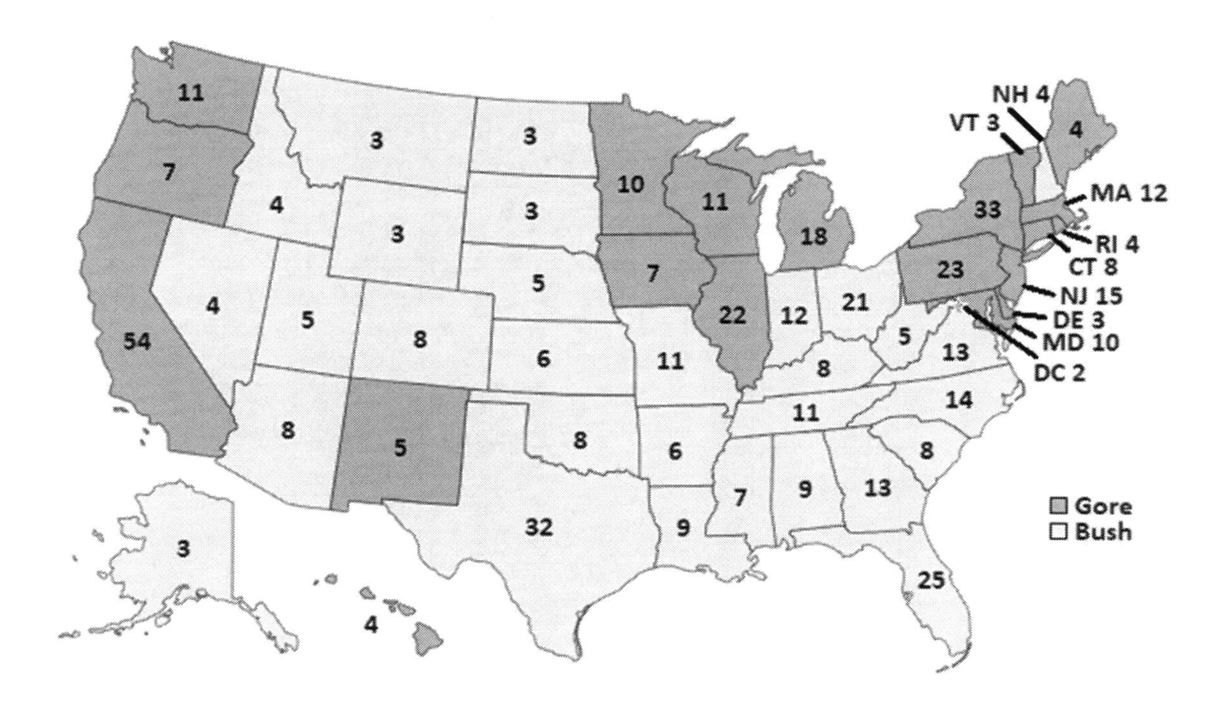

2000 Election Results			
Candidate	**Party**	**Electoral Votes**	**Popular Votes**
George W. Bush	Republican	271	50,456,062
Albert Gore, Jr.	Democratic	266	50,996,582

17. Based on the map, which of the following statements is NOT true?
    a. In 2000, the Democratic candidate won the popular vote, which means he won more the electoral votes of more states.
    b. In 2000, the Republican candidate won the presidential election because he won the electoral vote even though he did not win the popular vote.
    c. In 2000, the Democratic candidate won the popular vote but not the electoral vote, which reveals a discrepancy in the electoral system.
    d. In 2000, the Republican candidate won more states but did not win the popular vote.
    e. In 2000, the Republican candidate won 271 electoral votes, and the Democratic candidate won 266 electoral votes.

18. If you were asked to write a comprehensive research paper about life during the Great Depression in United States, which of the following would be a reliable primary source?
    a. Wikipedia article titled "Life in Depression America."
    b. Diary entry from Elsie May Long published in the article "The Great Depression: Two Kansas Diaries" by C. Robert Haywood in *Great Plains Quarterly*.
    c. Article titled "The Great Depression Begins: the Stock Market Crash of 1929," found at http://www.history.com/topics/great-depression
    d. Book by Glen H. Elder, Jr. titled *Children of the Great Depressions: Social Change in Life Experience*, published in 1999 by the American Psychological Association.
    e. Analysis of a diary entry titled "The Country's Depression: Oklahoma in a Time of Crisis."

19. "Excuse me, Sir," the host of the party declared, "this matter is not *any* of your business." The use of italics in this sentence indicates which of the following?

    a. Dialogue
    b. Thoughts
    c. Emphasis
    d. Volume
    e. A title

20. Snails, clams, mussels, and squid are mollusks, i.e., invertebrates with soft, unsegmented bodies that live in aquatic or damp habitats and have calcareous shells.

Which of the following does the abbreviation i.e. stand for in the preceding sentence?
    a. For example
    b. In error
    c. In addition
    d. That is
    e. Excepting

Questions 21–23 are based upon the following passage:

This excerpt is adaptation from Charles Dickens' speech in Birmingham in England on December 30, 1853 on behalf of the Birmingham and Midland Institute.

> My Good Friends,—When I first imparted to the committee of the projected Institute my particular wish that on one of the evenings of my readings here the main body of my audience should be composed of working men and their families, I was animated by two desires; first, by the wish to have the great pleasure of meeting you face to face at this Christmas time, and accompany you myself through one of my little Christmas books; and second, by the wish to have an opportunity of stating publicly in your presence, and in the presence of the committee, my earnest hope that the Institute will, from the beginning, recognise one great principle—strong in reason and justice—which I believe to be essential to the very life of such an Institution. It is, that the working man shall, from the first unto the last, have a share in the management of an Institution which is designed for his benefit, and which calls itself by his name.
>
> I have no fear here of being misunderstood—of being supposed to mean too much in this. If there ever was a time when any one class could of itself do much for its own good, and for the welfare of society—which I greatly doubt—that time is

unquestionably past. It is in the fusion of different classes, without confusion; in the bringing together of employers and employed; in the creating of a better common understanding among those whose interests are identical, who depend upon each other, who are vitally essential to each other, and who never can be in unnatural antagonism without deplorable results, that one of the chief principles of a Mechanics' Institution should consist. In this world a great deal of the bitterness among us arises from an imperfect understanding of one another. Erect in Birmingham a great Educational Institution, properly educational; educational of the feelings as well as of the reason; to which all orders of Birmingham men contribute; in which all orders of Birmingham men meet; wherein all orders of Birmingham men are faithfully represented—and you will erect a Temple of Concord here which will be a model edifice to the whole of England.

Contemplating as I do the existence of the Artisans' Committee, which not long ago considered the establishment of the Institute so sensibly, and supported it so heartily, I earnestly entreat the gentlemen—earnest I know in the good work, and who are now among us,—by all means to avoid the great shortcoming of similar institutions; and in asking the working man for his confidence, to set him the great example and give him theirs in return. You will judge for yourselves if I promise too much for the working man, when I say that he will stand by such an enterprise with the utmost of his patience, his perseverance, sense, and support; that I am sure he will need no charitable aid or condescending patronage; but will readily and cheerfully pay for the advantages which it confers; that he will prepare himself in individual cases where he feels that the adverse circumstances around him have rendered it necessary; in a word, that he will feel his responsibility like an honest man, and will most honestly and manfully discharge it. I now proceed to the pleasant task to which I assure you I have looked forward for a long time.

21. Based upon the contextual evidence provided in the passage above, what is the meaning of the term *enterprise* in the third paragraph?
   a. Company
   b. Courage
   c. Game
   d. Honesty
   e. Cause

22. The speaker addresses his audience as *My Good Friends*—what kind of credibility does this salutation give to the speaker?
   a. The speaker is an employer addressing his employees, so the salutation is a way for the boss to bridge the gap between himself and his employees.
   b. The speaker's salutation is one from an entertainer to his audience and uses the friendly language to connect to his audience before a serious speech.
   c. The salutation gives the serious speech that follows a somber tone, as it is used ironically.
   d. The speech is one from a politician to the public, so the salutation is used to grab the audience's attention.
   e. The salutation is one from a man on trial to his peers, so he is trying to evoke their empathy.

23. According to the aforementioned passage, what is the speaker's second desire for his time in front of the audience?
    a. To read a Christmas story
    b. For the working man to have a say in his institution which is designed for his benefit
    c. To have an opportunity to stand in their presence
    d. For the life of the institution to be essential to the audience as a whole
    e. To join them one day in the working forces.

Use the following image to answer questions 24 and 25.

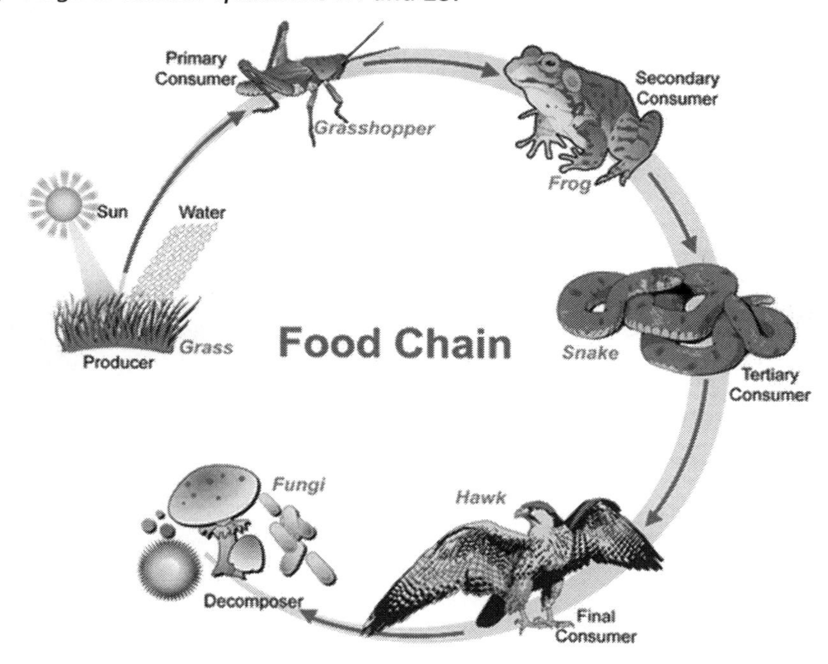

24. Which is the decomposer in the food chain above?
    a. Sun
    b. Grass
    c. Frog
    d. Fungi
    e. Snake

25. Which is the herbivore in the food chain above?
    a. Grass
    b. Hawk
    c. Frog
    d. Fungi
    e. Grasshopper

26. Consider the following headings that might be found under the entry for "Basketball" in an electronic encyclopedia.

- Introduction
- Basic Rules
- Professional Basketball
- College Basketball
- Similarities to Lacrosse
- Olympic and International Basketball
- Women's Basketball
- Bibliography

Which one does not belong?

a. Women's Basketball
b. Similarities to Lacrosse
c. Basic Rules
d. Bibliography
e. College Basketball

27. The guidewords at the top of a dictionary page are *receipt* and *reveal*. Which of the following words is NOT an entry on this page?

a. Receive
b. Retail
c. Revere
d. Reluctant
e. Reception

Use the nutrition label below for questions 28–30.

Nutrition Facts
Serving Size 2/3 cup (55g)
Servings Per Container About 8

Amount Per Serving

Calories 230	Calories from Fat 72
	% Daily Value*
Total Fat 8g	**12%**
Saturated Fat 1g	**5%**
Trans Fat 0g	
Cholesterol 0mg	**0%**
Sodium 160mg	**7%**
Total Carbohydrate 37g	**12%**
Dietary Fiber 4g	**16%**
Sugars 1g	
Protein 3g	
Vitamin A	10%
Vitamin C	8%
Calcium	20%
Iron	45%

* Percent Daily Values are based on a 2,000 calorie diet. Your daily value may be higher or lower depending on your calorie needs.

	Calories:	2,000	2,500
Total Fat	Less than	65g	80g
Sat Fat	Less than	20g	25g
Cholesterol	Less than	300mg	300mg
Sodium	Less than	2,400mg	2,400mg
Total Carbohydrate		300g	375g
Dietary Fiber		25g	30g

28. Based on the information provided on the nutrition label, approximately what percent of the total calories come from fat?
   a. 12%
   b. 33%
   c. 5%
   d. 0%
   e. 25%

29. If Alex ate two cups of this product, approximately how many calories did he consume?
   a. Almost 700 calories
   b. A little more than 200 calories
   c. Around 450 calories
   d. A little more than 70
   e. Around 300 calories

30. Based on the information from the nutrition label above, which of the following statements is true?
   a. Someone eating this food item would not need to worry for the rest of the day about getting more of vitamins A and C.
   b. This food item is not a good source of iron.
   c. For someone consuming 2,000 calories a day, this food item contains too much dietary fiber.
   d. Someone on a low sodium diet would not need to worry about this food item being high in sodium.
   e. The trans-fat found in this food would exceed the appropriate amounts per day.

31. In her famous poem "Because I Could Not Stop for Death," Emily Dickinson writes, "Because I could not stop for Death / He kindly stopped for me." Her mention of death is an example of which of the following?
   a. Hyperbole
   b. Personification
   c. Allusion
   d. Alliteration
   e. Simile

32. In the poem quoted above, what does the slash mark between the words "Death" and "He" represent?
   a. The break between two stanzas
   b. The end of one sentence and the start of another
   c. A punctuation mark that stands in for a comma
   d. The shift from one speaker to another
   e. The break between two lines of poetry

Use the following three quotations from Thomas Jefferson to answer questions 33 and 34.

> "The tree of liberty must be refreshed from time to time with the blood of patriots and tyrants."
>
> "In matters of style, swim with the current; in matters of principle, stand like a rock."
>
> "I hold it, that a little rebellion, now and then, is a good thing, and as necessary in the political world as storms in the physical."

33. What is Jefferson's opinion on conflict?
   a. It is unavoidable and necessary.
   b. It must be avoided at all costs.
   c. Staying true to your principles is never worth the price you might pay.
   d. It will lead to ultimate destruction.
   e. It is unnecessary though not avoidable.

34. The "tree of liberty" is an example of which of the following?
   a. Personification
   b. Allusion
   c. Analogy
   d. Idiom
   e. Hyperbole

35. If you were looking for the most reliable and up-to-date information regarding the safety of travel overseas, which of the following websites would be the most accurate?
   a. http://www.nomadicmatt.com/travel-blog/
   b. https://en.wikipedia.org/wiki/Tourism
   c. http://gizmodo.com/how-to-travel-internationally-for-the-very-first-time
   d. http://www.state.gov/travel/
   e. https://travel.usnews.com/rankings/usa/

Use the following excerpt for questions 36 and 37.

> Evidently, our country has overlooked both the importance of learning history and the appreciation of it. But why is this a huge problem? Other than historians, who really cares how the War of 1812 began, or who Alexander the Great's tutor was? Well, not many, as it turns out. So, *is* history really that important? Yes! History is *critical* to help us understand the underlying forces that shape decisive events, to prevent us from making the same mistakes twice, and to give us context for current events.

36. The above is an example of which type of writing?
   a. Expository
   b. Persuasive
   c. Narrative
   d. Poetry
   e. Technical

37. In the above passage, what does the use of boldface type for the word "critical" indicate?
   a. Dialogue
   b. Volume
   c. Emphasis
   d. Misspelling
   e. Inaccurate information

38. The guidewords at the top of a dictionary page are *able-bodied* and *about-face*. Which of the following words appear as an entry on this page?
   a. Abolition
   b. Abundant
   c. Able
   d. Ability
   e. Abuse

39. Synonyms of the word *spurious* are *feigned* and *fraudulent*. Which of the following is an antonym for *spurious*?
   a. Phony
   b. Impersonated
   c. Misleading
   d. Genuine
   e. Contrived

40. Which of the following sentences contains an opinion statement?
    a. In 1819, the Supreme Court ruled that Congress could create a national bank.
    b. Marshall also ruled that the states did not have the right to tax the bank or any other agency created by the federal government.
    c. John Marshall was one of the most intelligent chief justices in the United States.
    d. The chief justice is the head of the judicial branch of the government.
    e. John Marshall was the fourth Chief Justice of the United States.

41. Which of the following sentences is a factual statement?
    a. Many nutritionists believe a low carbohydrate, high protein diet is the healthiest diet.
    b. Legislation should be passed mandating that cell phones be banned in all public-school classrooms.
    c. Spanish is an easier language to learn than Japanese.
    d. College students would benefit greatly from participating in intramural sports on their campuses.
    e. Macbook computers are easier to use than PC.

42. Follow the numbered instructions to transform the starting word into a different word.

    1. Start with the word *CARLOAD*.
    2. Move the first letter to the end of the word.
    3. Switch the first and second letters.
    4. Switch the third and sixth letters.
    5. Change the fourth letter to an *i*.
    6. Move the last letter before the second *a*.

What is the new word?
    a. Carpool
    b. Radical
    c. Ordeal
    d. Railroad
    e. Rainbow

Questions 43 and 44 refer to the following quote from Martin Luther King Jr.'s Nobel Peace Prize acceptance speech in Oslo on December 10, 1964.

> "I refuse to accept the view that mankind is so tragically bound to the starless midnight of racism and war that the bright daybreak of peace and brotherhood can never become a reality . . . I believe that unarmed truth and unconditional love will have the final word."

43. Which of the following statements is NOT accurate based on this quote?
    a. Martin Luther King Jr. felt that the fight against racism was ultimately hopeless due to mankind's primitive instincts.
    b. Martin Luther King Jr. was eternally optimistic about mankind's ability to overcome racism.
    c. Martin Luther King Jr. believed that the war against racism could be won with truth and love.
    d. Martin Luther King Jr. believed in that the goodness of mankind would prevail.
    e. Martin Luther King Jr. felt that "bright daybreak" was a proper metaphor for peace and brotherhood.

44. By referring to the "starless midnight of racism," Martin Luther King means:
    a. Racism involves prejudices against people with dark skin.
    b. Racism is evil and blinds people to the truth.
    c. It is easier to be racist on a starless night.
    d. The problem of racism is not as bad during the day.
    e. Racism has historically been associated with the moon.

Questions 45–48 refer to the following passage.

> Although many Missourians know that Harry S. Truman and Walt Disney hailed from their great state, probably far fewer know that it was also home to the remarkable George Washington Carver. At the end of the Civil War, Moses Carver, the slave owner who owned George's parents, decided to keep George and his brother and raise them on his farm. As a child, George was driven to learn and he loved painting. He even went on to study art while in college but was encouraged to pursue botany instead. He spent much of his life helping others by showing them better ways to farm; his ideas improved agricultural productivity in many countries. One of his most notable contributions to the newly emerging class of Negro farmers was to teach them the negative effects of agricultural monoculture, i.e. growing the same crops in the same fields year after year, depleting the soil of much needed nutrients and resulting in a lesser yielding crop. Carver was an innovator, always thinking of new and better ways to do things, and is most famous for his over three hundred uses for the peanut. Toward the end of his career, Carver returned to his first love of art. Through his artwork, he hoped to inspire people to see the beauty around them and to do great things themselves. When Carver died, he left his money to help fund ongoing agricultural research. Today, people still visit and study at the George Washington Carver Foundation at Tuskegee Institute.

45. Which of the following describes the kind of writing used in the above passage?
    a. Narrative
    b. Persuasive
    c. Technical
    d. Expository
    e. Poetry

46. According to the passage, what was George Washington Carver's first love?
    a. Plants
    b. Music
    c. Animals
    d. Soil
    e. Art

47. According to the passage, what is the best definition for agricultural monoculture?
    a. The practice of producing or growing a single crop or plant species over a wide area and for a large number of consecutive years
    b. The practice of growing a diversity of crops and rotating them from year to year
    c. The practice of growing crops organically to avoid the use of pesticides
    d. The practice of charging an inflated price for cheap crops to obtain a greater profit margin
    e. A place where farmers come together and sell their goods to the public

48. Which of the following is the best summary of this passage?
 a. George Washington Carver was born at a time when scientific discovery was at a virtual standstill.
 b. Because he was African American, there were not many opportunities for George Washington Carver.
 c. George Washington Carver was an intelligent man whose research and discoveries had an impact worldwide.
 d. George Washington Carver was far more successful as an artist than he was as a scientist.
 e. George Washington Carver placed duty above passion for the good of the people.

Questions 49–50 are based upon the following passage:

This excerpt is adaptation from *The Life-Story of Insects,* by Geo H. Carpenter.

> Insects as a whole are preeminently creatures of the land and the air. This is shown not only by the possession of wings by a vast majority of the class, but by the mode of breathing to which reference has already been made, a system of branching air-tubes carrying atmospheric air with its combustion-supporting oxygen to all the insect's tissues. The air gains access to these tubes through a number of paired air-holes or spiracles, arranged segmentally in series.
>
> It is of great interest to find that, nevertheless, a number of insects spend much of their time under water. This is true of not a few in the perfect winged state, as for example aquatic beetles and water-bugs ('boatmen' and 'scorpions') which have some way of protecting their spiracles when submerged, and, possessing usually the power of flight, can pass on occasion from pond or stream to upper air. But it is advisable in connection with our present subject to dwell especially on some insects that remain continually under water till they are ready to undergo their final moult and attain the winged state, which they pass entirely in the air. The preparatory instars of such insects are aquatic; the adult instar is aerial. All may-flies, dragon-flies, and caddis-flies, many beetles and two-winged flies, and a few moths thus divide their life-story between the water and the air. For the present we confine attention to the Stone-flies, the May-flies, and the Dragon-flies, three well-known orders of insects respectively called by systematists the Plecoptera, the Ephemeroptera and the Odonata.
>
> In the case of many insects that have aquatic larvae, the latter are provided with some arrangement for enabling them to reach atmospheric air through the surface-film of the water. But the larva of a stone-fly, a dragon-fly, or a may-fly is adapted more completely than these for aquatic life; it can, by means of gills of some kind, breathe the air dissolved in water.

49. Which statement best details the central idea in this passage?
 a. It introduces certain insects that transition from water to air.
 b. It delves into entomology, especially where gills are concerned.
 c. It defines what constitutes as insects' breathing.
 d. It invites readers to have a hand in the preservation of insects.
 e. It explains the life expectancy of the boatman and the scorpion.

50. Which definition most closely relates to the usage of the word *moult* in the passage?
   a. An adventure of sorts, especially underwater
   b. Mating act between two insects
   c. The act of shedding part or all of the outer shell
   d. Death of an organism that ends in a revival of life
   e. The last stage before the insect becomes an adult

Mathematics

1. If $6t + 4 = 16$, what is t?
   a. 1
   b. 2 ✓
   c. 3
   d. 4
   e. 5

2. The variable *y* is directly proportional to *x*. If $y = 3$ when $x = 5$, then what is *y* when $x = 20$?
   a. 10
   b. 12
   c. 14
   d. 16
   e. 18

3. A line passes through the point (1, 2) and crosses the *y*-axis at *y* = 1. Which of the following is an equation for this line?
   a. $y = 2x$
   b. $y = 3x$
   c. $x + y = 1$
   d. $y = \frac{x}{2} - 2$
   e. $y = x + 1$

4. There are $4x + 1$ treats in each party favor bag. If a total of $60x + 15$ treats are distributed, how many bags are given out?
   a. 15
   b. 16
   c. 20
   d. 22
   e. 25

5. Apples cost $2 each, while bananas cost $3 each. Maria purchased 10 fruits in total and spent $22. How many apples did she buy?
   a. 5
   b. 6
   c. 7
   d. 8 ✓
   e. 9

6. What are the polynomial roots of $x^2 + x - 2$?
   a. 1 and -2
   b. -1 and 2
   c. 2 and -2
   d. 9 and 13
   e. 4 and 13

7. What is the *y*-intercept of $y = x^{5/3} + (x - 3)(x + 1)$?
   a. 3.5
   b. 7.6
   c. -3
   d. -15.1
   e. -5.5

8. $x^4 - 16$ can be simplified to which of the following?
   a. $(x^2 + 2)(x^2 - 4)$
   b. $(x^2 + 4)(x^2 + 4)$
   c. $(x^2 - 4)(x^2 - 4)$
   d. $(x^2 - 2)(x^2 + 4)$
   e. $(x^2 - 4)(x^2 + 4)$

9. $(4x^2y^4)^{\frac{3}{2}}$ can be simplified to which of the following?
   a. $8x^3y^6$
   b. $4x^{\frac{5}{2}}y$
   c. $4xy$
   d. $32x^{\frac{7}{2}}y^{\frac{11}{2}}$
   e. $8x^2y^3$

10. If $\sqrt{1 + x} = 4$, what is *x*?
   a. 10
   b. 15
   c. 20
   d. 25
   e. 30

11. Suppose $\frac{x+2}{x} = 2$. What is *x*?
   a. -1
   b. 0
   c. 2
   d. 4
   e. 5

12. A ball is thrown from the top of a high hill, so that the height of the ball as a function of time is $h(t) = -16t^2 + 4t + 6$, in feet. What is the maximum height of the ball in feet?

a. 6
b. 6.25
c. 6.5
d. 6.75
e. 7.25

13. A rectangle has a length that is 5 feet longer than three times its width. If the perimeter is 90 feet, what is the length in feet?

a. 10
b. 20
c. 25
d. 35
e. 45

14. Five students take a test. The scores of the first four students are 80, 85, 75, and 60. If the median score is 80, which of the following could NOT be the score of the fifth student?

a. 60
b. 80
c. 85
d. 100
e. 110

15. In an office, there are 50 workers. A total of 60% of the workers are women, and the chances of a woman wearing a skirt is 50%. If no men wear skirts, how many workers are wearing skirts?

a. 12
b. 15
c. 16
d. 20
e. 22

16. Ten students take a test. Five students get a 50. Four students get a 70. If the average score is 55, what was the last student's score?

a. 20
b. 40
c. 50
d. 60
e. 70

17. A company invests $50,000 in a building where they can produce saws. If the cost of producing one saw is $40, then which function expresses the amount of money the company pays? The variable *y* is the money paid and *x* is the number of saws produced.

a. $y = 50{,}000x + 40$
b. $y + 40 = x - 50{,}000$
c. $y = 40x - 50{,}000$
d. $y = 50x - 400{,}000$
e. $y = 40x + 50{,}000$

18. A six-sided die is rolled. What is the probability that the roll is 1 or 2?

a. $\frac{1}{6}$

b. $\frac{1}{4}$

c. $\frac{1}{3}$

d. $\frac{1}{2}$

e. $\frac{5}{2}$

19. A line passes through the origin and through the point (-3, 4). What is the slope of the line?

a. $-\frac{4}{3}$

b. $-\frac{3}{4}$

c. $\frac{4}{3}$

d. $\frac{3}{4}$

e. $-\frac{1}{2}$

20. An equilateral triangle has a perimeter of 18 feet. If a square whose sides have the same length as one side of the triangle is built, what will be the area of the square?

a. 6 square feet
b. 36 square feet
c. 256 square feet
d. 1000 square feet
e. 1200 square feet

21. Change $3\frac{3}{5}$ to a decimal.

a. 3.6
b. 4.67
c. 5.3
d. 0.28
e. 1.77

22. If a car can travel 300 miles in 4 hours, how far can it go in an hour and a half?

a. 100 miles
b. 112.5 miles
c. 135.5 miles
d. 150 miles
e. 223 miles

23. Which measure for the center of a small sample set is most affected by outliers?

a. Mean
b. Median
c. Mode
d. Range
e. None of the above

24. Given the value of a given stock at monthly intervals, which graph should be used to best represent the trend of the stock?

a. Box plot
b. Line plot
c. Line graph
d. Circle graph
e. Pie chart

25. What is the probability of randomly picking the winner and runner-up from a race of 4 horses and distinguishing which is the winner?

a. $\frac{1}{4}$

b. $\frac{1}{2}$

c. $\frac{1}{16}$

d. $\frac{1}{12}$

e. $\frac{1}{15}$

26. Which of the following could be used in the classroom to show $\frac{3}{7} < \frac{5}{6}$ is a true statement?

a. A bar graph
b. A number line
c. An area model
d. Base 10 blocks
e. A line graph

27. Add $103{,}678 + 487$

a. 103,191
b. 103,550
c. 104,265
d. 104,165
e. 105,270

28. Add $1.001 + 5.629$

a. 4.472
b. 4.628
c. 5.630
d. 6.628
e. 6.630

29. Add $143.77 + 5.2$
   a. 138.57
   b. 148.97
   c. 138.97
   d. 148.57
   e. 149.67

30. What is the next number in the following series: $1, 3, 6, 10, 15, 21, \dots$?
   a. 26
   b. 27
   c. 28
   d. 29
   e. 31

31. Add and express in reduced form $\frac{14}{33} + \frac{10}{11}$.
   a. $\frac{2}{11}$
   b. $\frac{6}{11}$
   c. $\frac{4}{3}$
   d. $\frac{44}{33}$
   e. $\frac{1}{120}$

32. 32 is 25% of what number?
   a. 64
   b. 128
   c. 12.65
   d. 8
   e. 15

33. Which of the following is NOT a way to write 40 percent of *N*?
   a. $(0.4)N$
   b. $\frac{2}{5}N$
   c. $40N$
   d. $\frac{4N}{10}$
   e. $\frac{8}{20}N$

34. Subtract $112{,}076 - 1{,}243$.
   a. 109,398
   b. 113,319
   c. 113,833
   d. 110,319
   e. 110,833

35. Carey bought 184 pounds of fertilizer to use on her lawn. Each segment of her lawn required $12\frac{1}{2}$ pounds of fertilizer to do a sufficient job. If a student were asked to determine how many segments could be fertilized with the amount purchased, what operation would be necessary to solve this problem?

a. Multiplication
b. Division
c. Addition
d. Subtraction
e. Exponentiation

36. Subtract $701.1 - 52.33$.

a. 753.43
b. 648.77
c. 652.77
d. 638.43
e. 524.87

37. Which of the following expressions best exemplifies the additive and subtractive identity?

a. $5 + 2 - 0 = 5 + 2 + 0$
b. $6 + x = 6 - 6$
c. $9 - 9 = 0$
d. $8 + 2 = 10$
e. $7 + 2 = 8x - 9$

38. Which of the following is an equivalent measurement for 1.3 cm?

a. 0.13 m
b. 0.013 m
c. 0.13 mm
d. 0.013 mm
e. 1.3 mm

39. Divide $1{,}015 \div 1.4$.

a. 7,250
b. 0.725
c. 7.25
d. 72.50
e. 725

40. At the store, Jan spends $90 on apples and oranges. Apples cost $1 each and oranges cost $2 each. If Jan buys the same number of apples as oranges, how many oranges did she buy?

a. 20
b. 25
c. 30
d. 35
e. 40

41. Multiply 12.4×0.2.
  a. 12.6
  b. 2.48
  c. 12.48
  d. 2.6
  e. 18.2

42. Multiply $1{,}987 \times 0.05$.
  a. 9.935
  b. 99.35
  c. 993.5
  d. 999.35
  e. 87.99

43. Which item taught in the classroom would allow students to correctly find the solution to the following problem: A clock reads 5:00 am. What is the measure of the angle formed by the two hands of that clock?
  a. Each time increment on an analog clock measures 90 degrees.
  b. Each time increment on an analog clock measures 30 degrees.
  c. Two adjacent angles sum up to 180 degrees.
  d. Two complementary angles sum up to 180 degrees.
  e. Two complementary angles sum up to 90 degrees.

44. Divide, express with a remainder $1{,}202 \div 44$.
  a. $27\frac{2}{7}$
  b. $2\frac{7}{22}$
  c. $7\frac{2}{7}$
  d. $27\frac{7}{22}$
  e. $9\frac{2}{27}$

45. What is the volume of a box with rectangular sides 5 feet long, 6 feet wide, and 3 feet high?
  a. 60 cubic feet
  b. 75 cubic feet
  c. 90 cubic feet
  d. 14 cubic feet
  e. 5 cubic feet

46. Divide $702 \div 2.6$.
  a. 27
  b. 207
  c. 2.7
  d. 3.7
  e. 270

47. A train traveling 50 miles per hour takes a trip lasting 3 hours. If a map has a scale of 1 inch per 10 miles, how many inches apart are the train's starting point and ending point on the map?
    a. 14
    b. 12
    c. 13
    d. 15
    e. 17

48. A traveler takes an hour to drive to a museum, spends 3 hours and 30 minutes there, and takes half an hour to drive home. What percentage of his or her time was spent driving?
    a. 15%
    b. 30%
    c. 40%
    d. 60%
    e. 65%

49. A truck is carrying three cylindrical barrels. Their bases have a diameter of 2 feet and they have a height of 3 feet. What is the total volume of the three barrels in cubic feet?
    a. 3π
    b. 9π
    c. 12π
    d. 15π
    e. 20π

50. Greg buys a $10 lunch with 5% sales tax. He leaves a $2 tip after his bill. How much money does he spend?
    a. $12.50
    b. $12
    c. $13
    d. $13.25
    e. $11.50

Writing

First Essay Prompt

Directions: The Writing section of the CBEST will require test takers to write two essays based on provided prompts. The first essay has a referential aim so that test takers can showcase their analytical and expository writing skills. The second essay will have an expressive aim, with a topic relating to the test taker's past lived experience.

Please read the prompt below and answer in an essay format.

> Albert Einstein said that "Everybody is a genius. But if you judge a fish by its ability to climb a tree, it will live its whole life believing that it is stupid." What do you think Einstein meant by this statement? Support your answer with details and observations.

Second Essay Prompt

Please read the prompt below and answer in an essay format.

> Write an essay detailing your experience with loss and how you dealt with it. Talk about the process of the loss and if or how it changed you. How do you view life today in light of that loss?

Answer Explanations #1

Reading

1. C: All the details in this paragraph suggest that Brookside is a great place to live, plus the last sentence states that it is an *idyllic neighborhood*, meaning it is perfect, happy, and blissful. Choices *A* and *B* are incorrect, because although they do contain specific details from the paragraph that support the main idea, they are not the main idea. Choice *D* is incorrect because there is no reference in the paragraph to the crime rate in Brookside. Choice *E* is incorrect; the author does think Brookside is a great place, but they don't try and convince the audience to move there.

2. E: A passage like this one would likely appear in some sort of community profile, highlighting the benefits of living or working there. Choice *A* is incorrect because nothing in this passage suggests that it is fictional. It reads as non-fiction, if anything. Choice *B* is incorrect; it is more an informational passage than a creative passage. Choice *C* is incorrect because it does not report anything particularly newsworthy, and Choice *D* is incorrect because it has absolutely nothing to do with a movie review.

3. D: In the first sentence, it states very clearly that the Brookside neighborhood is *one of the nation's first planned communities*. This makes it unique, as many other neighborhoods are old, many other neighborhoods exist in Kansas City, and many other neighborhoods have shopping areas. For these reasons, all the other answer choices are incorrect.

4. B: Chapter 4, Section II would have the information for a research paper about how the nation plans to address the problem of child neglect in affected communities. The title of this section is called "Components of the Commission's National Strategy."

5. A: The chapter on page 52 is called "Addressing the Needs of American Indian/Alaska Native Children." Choice *B* is the chapter related to child abuse in disproportionately affected communities. Choice *C* is related to law enforcement and CPS. Choice *D* is related to support for families. Choice *E* is related to a solution or recommendation for the problems. Thus, Choice *A* is the correct answer choice.

6. C: Child abuse and neglect is alive and rampant in our communities, and this chapter outlines those who are in dire need of our attention. Choice *A* has more to do with Chapter 2, "Saving Children's Lives Today and Into the Future," and thus is incorrect. Choice *B* relates to Chapter 4, "Reducing Child Abuse and Neglect Deaths in Disproportionately Affected Communities," and is also incorrect. Choice *D* relates to Chapter 7, "Multidisciplinary Support for Families," which is incorrect. Choice *E* relates to Chapter 6, "Decisions Grounded in Better Data and Research."

7. E: Choice *E* is correct because a *simile* is a comparison between two unlike things using the words *like* or *as*. Choice *A* is incorrect because *hyperbole* is an exaggeration for affect. Choice *B* is incorrect; since the simile uses "like," the comparison cannot be considered a metaphor. Choice *C* is incorrect because *personification* attributes human characteristics to an inanimate object. Choice *D* is incorrect because *alliteration* is a poetic device in which several words begin with the same letter or letters. It is used to add emphasis or affect.

8. A: Choice *A* is correct because there is evidence in the passage to support it, specifically when he mentions catching "a mess of black-fish, which you couldn't buy in New York for a dollar—large fat fellows, with meat on their bones that it takes a pretty long fork to stick through." Choice *E* is a detail at the beginning of the passage, but it's not a summary of the passage as a whole.

9. B: Prodigious means *great in extent, size, or degree*. In this passage, Whitman is comparing the easternmost part of Long Island to a very large alligator. Therefore, the other answer choices are incorrect.

10. C: Choice *C* is correct because it states the fact that education has a positive impact on lowering crime rates, and it tells your reader what you will cover in that paragraph without providing specific details yet. That's what the remainder of the paragraph is for. Choice *A* is incorrect because the paragraph is about the verification of crime statistics. Choice *B* is incorrect because it is a broad generalization that is simply not true. Choice *D*, while it is on topic, is too specific to be a topic sentence but would be an excellent supporting detail. Choice *E* acts as the beginning of a counterargument so this cannot be the topic sentence.

11. E: This sentence would not be a good thesis statement because it conveys one very specific detail. All the other answer choices would be good thesis statements because they are general enough for an entire essay but specific enough so that the reader knows exactly what the essay will cover.

12. B: The author clearly states that education is crucial to the survival of the human race, and it can be easily inferred that if this is true, then improvements to our educational system are certainly worth fighting for. Choices *A* and *C* are incorrect because there is nothing in the passage that relates to these statements. Choice *D* is incorrect because it directly contradicts what the author states about all children's ability to learn. Choice *E* is mentioned in the passage, but it is not the main point.

13. A: Clearly, this author feels passionately about the importance of education. This is evident especially in the word choices. For this reason, all the other answer choices are incorrect.

14. C: Based on the author's passionate stance about the importance of education for all children, this answer choice makes the most sense. For this reason, all the other answer choices are incorrect.

15. D: The author mentions the importance of parent involvement and communication between school and home. He also devotes one full paragraph to the impact of technology on education. Nowhere in the passage does the author mention the cost of textbooks, so Choice *D* is correct.

16. C: In 1964, the Democratic candidate won 486 electoral votes while the Republican candidate only won 52, and in 1984, almost the complete opposite was true—the Republican candidate won 525 electoral votes, while the Democratic candidate only won 13. Clearly, this was an almost total swing from one party to the other. Choices *A* and *B* are incorrect because the voter turnout, as indicated by the popular vote numbers, did not remain the same or decrease over this twenty-year period. The popular vote turnout increased by approximately twenty thousand over this twenty-year period. Choice *D* is incorrect because the electoral votes won by each party did change during this twenty-year period. Choice *E* is incorrect; this is the opposite of the correct answer.

17. A: The Democratic candidate did win the popular vote, but he lost the electoral votes by five, and a closer look at the map reveals that he only won 21 out of the possible 50 states. All the other statements are correct.

18. B: Primary sources are original, first-hand accounts of events or time-periods as they are happening or very close to the time they occurred. Diary entries are excellent primary sources, as are newspaper articles, works of art and literature, interviews, and live recordings. Choice *A* is incorrect because a Wikipedia article is not a reliable source due to the fact that multiple authors can access and manipulate the information. Choices *C* and *D* are incorrect because they are examples of secondary sources. While they might be very reliable and useful, they are not primary sources. Choice *E* describes an analysis of a primary source, which is a secondary source, so this is incorrect.

19. C: Italics are often used to indicate emphasis. Choice *A* is incorrect because quotation marks, not italics, are used to indicate dialogue. Choice *B* is incorrect because although italics can also indicate someone's interior thoughts, in this case, it would not make sense for the word *any* to be a thought rather than spoken aloud. Choice *D* is incorrect because volume is typically indicated by capital letters. Choice *E* is incorrect; sometimes the use of italics indicates a title of a film or novel, but the context here tells us that someone is emphasizing speech rather than giving us a title.

20. D: The abbreviation *i.e.* stands for the Latin phrase, "id est" meaning "that is." It is used to clarify an idea, in this case, the term "mollusk." Choice *A* is incorrect because the abbreviation meaning "for example" is *e.g.* Choices *B, C,* and *E* are incorrect because there are no abbreviations in English for these phrases.

21. E: *Enterprise* most closely means *cause*. Choices *A, B,* and *C* are all related to the term *enterprise*. However, Dickens speaks of a *cause* here, not a company, courage, or a game. *He will stand by such an enterprise* is a call to stand by a cause to enable the working man to have a certain autonomy over his own economic standing. The very first paragraph ends with the statement that the working man *shall . . . have a share in the management of an institution which is designed for his benefit.*

22. B: The speaker's salutation is one from an entertainer to his audience and uses the friendly language to connect to his audience before a serious speech. Recall in the first paragraph that the speaker is there to "accompany [the audience] . . . through one of my little Christmas books," making him an author there to entertain the crowd with his own writing. The speech preceding the reading is the passage itself, and, as the tone indicates, a serious speech addressing the "working man." Although the passage speaks of employers and employees, the speaker himself is not an employer of the audience, so Choice *A* is incorrect. Choice *C* is also incorrect, as the salutation is not used ironically, but sincerely, as the speech addresses the wellbeing of the crowd. Choice *D* is incorrect because the speech is not given by a politician, but by a writer. Choice *E* is incorrect, as there is no indication the speaker is on trial.

23. B: For the working man to have a say in his institution which is designed for his benefit Choice *A* is incorrect because that is the speaker's *first* desire, not his second. Choices *C* and *D* are tricky because the language of both of these is mentioned after the word *second*. However, the speaker doesn't get to the second wish until the next sentence. Choices *C* and *D* are merely prepositions preparing for the statement of the main clause, Choice *B*. Choice *E* is not in the passage.

24. D: Fungi. Choice *A* (the sun) is not even a living thing. Grass (*B*) is a producer, and the frog (*C*) and snake (*E*) are consumers. The fungi break down dead organisms and are the only decomposer shown.

25. E: Grasshopper. An herbivore is an organism that eats only plants, and that's the grasshopper's niche in this particular food chain. Grass (*A*) is a producer, the frog and hawk (*C* and *E*) are consumers, and the fungi (*D*) is a decomposer.

26. B: While lacrosse does develop some similar skill sets as basketball, it does not belong in an encyclopedia entry about basketball. All the other answer choices do contain relevant information that would likely appear in an encyclopedia entry under "Basketball."

27. C: The word *revere* would not be an entry on this page because the letters *rever* would come after *revea* in alphabetical order. All the other answer choices contain letters that would come between receipt and reveal and would therefore appear as entries on this page.

28. B: According to the label, 72 calories out of the total 230 calories come from fat. This is approximately a third of 230, or 33%. Choice *A* is incorrect because 12% is the percentage of the daily value of total fat found in a serving of this product. Choice *C* is incorrect because 5% is the percentage of the daily value of saturated fat found in a serving of this product. Choice *D* is incorrect because 0% represents the number of grams of trans fat found in a serving of this product. Choice *E* is incorrect because it is just a random number.

29. A: This is the correct answer because if Alex ate two cups, that would be exactly three full servings, based on the label information that 2/3 cup is one serving size. One serving contains 230 calories, so three servings would be three times that amount, or 690 calories. Choice *B* is incorrect because that reflects the number of calories in one serving. Choice *C* is incorrect because that reflects the number of calories in two servings. Choice *D* is incorrect because that reflects the number of calories from fat in one serving. Choice *E* does not equal a certain amount of servings here.

30. D: This is the correct answer because this product only contains 160 mg, or 7% of the daily value of sodium, so this item would be perfectly fine for someone on a low sodium diet to consume. Choice *A* is incorrect because the values of vitamins A and C are relatively low, meaning someone should try to consume more of those throughout the day. Choice *B* is incorrect because this item contains 45% of the daily value of iron, so it is a good source of iron. Choice *C* is incorrect because this item is not too high in dietary fiber for someone consuming 2,000 calories a day. Choice *E* is incorrect; there is no trans-fat in this food.

31. B: The correct answer is *personification* because Dickinson gives death characteristics of a human person. In addition, by capitalizing the word, she has made "Death" a proper noun or name. Choice *A* is incorrect because *hyperbole* is an exaggeration for affect. Choice *C* is incorrect because an *allusion* is a brief or indirect reference to a person, place, or thing from history, culture, literature, or politics. Choice *D* is incorrect because *alliteration* is a poetic device in which several words begin with the same letter or letters. It is used to add emphasis or affect. Choice *E* is incorrect because a simile compares two things using the words "like" or "as."

32. E: This is the correct answer because the slash mark represents the break between the first line of this poem and the second. A slash mark would not be used to indicate any of the other answer choices.

33. A: This is the correct answer because he makes it clear that sometimes conflict must happen, and that it can even be a good thing.

34. C: Analogy is the correct answer because an analogy is a comparison between two seemingly unlike things, used to help clarify meaning. In this case, Jefferson is comparing what it takes to keep the ideal of liberty alive to a tree. Choice *A* is incorrect because *personification* attributes human traits or characteristics to an inanimate object. Choice *B* is incorrect because an *allusion* is a brief or indirect reference to a person, place, or thing from history, culture, literature, or politics. Choice *D* is incorrect because an *idiom* is an expression whose meaning is not to be taken literally. Choice *E*, hyperbole, is an exaggeration.

35. D: This is the correct answer because the "gov" indicates that this is a government website. If you are traveling overseas, the U.S. government would be a reliable and up-to-date source of information, especially in regard to travel safety. While the other websites may contain some helpful information, and may indeed be worth reading, blogs and wikis may not be as reliable.

36. B: Persuasive is the correct answer because the author is clearly trying to convey the point that history education is very important. Choice *A* is incorrect because expository writing is more informational and less emotional. Choice *C* is incorrect because narrative writing involves story telling. Choice *D* is incorrect because this is a piece of prose, not poetry. Choice *E* is incorrect because technical writing represents instruction manuals and proposals and is mostly neutral and objective.

37. C: This is the correct answer because boldface type is often used to indicate emphasis.

Choice *A* is incorrect because quotation marks are used to indicate dialogue. Choice *B* is incorrect because volume is typically indicated by the use of capital letters. Choices *D* and *E* are incorrect because there is no way to indicate misspelling or inaccurate information by altering the text.

38. A: Choice *A* is correct because the letters *abol* fall between the two guidewords in alphabetical order. All the other answer choices are incorrect because they do not fall between the two guidewords in alphabetical order.

39. D: An antonym is a word that means the opposite, so *genuine* is the antonym of *feigned,* which means falsified or pretended. All the other answer choices are synonyms for the word *feigned*.

40. C: This answer choice is correct because *one of the most intelligent* is a matter of opinion, not a quantifiable fact. All the other answer choices are factual statements.

41. A: This answer choice is correct because it is the only statement that is not based on opinion. Nutritionists' belief in one certain diet *might* be a matter of opinion, but the statement that many do believe in the health benefits of this particular diet is a fact.

42. B: Following the steps in order will change the word *carload* into the word *radical*.

43. A: This is the correct answer choice because there is evidence to the contrary that Martin Luther King Jr. refused to believe that mankind would be bound to the darkness of racism.

44. B: This is the correct answer choice because Martin Luther King Jr. is using an analogy, comparing the evil of racism to the darkness one would experience on a starless night. He is also making the point that just as one cannot see in the dark, racism blinds one to the truth.

45. D: This is the correct answer choice because expository writing involves straightforward, factual information and analysis. It is unbiased and does not rely on the writer's personal feelings or opinions. Choice *A* is incorrect because narrative writing tells a story. Choice *B* is incorrect because persuasive writing is intended to change the reader's mind or position on a topic. Choice *C* is incorrect because technical writing attempts to outline a complex object or process. Choice *E* is incorrect because poetry usually has line breaks within stanzas, and the formatting usually does not look like a single paragraph with objective information.

46. E: This is the correct answer choice because the passage begins by describing Carver's childhood fascination with painting and later returns to this point when it states that at the end of his career "Carver returned to his first love of art." For this reason, all the other answer choices are incorrect.

47. A: This is the correct answer choice because the passage contains a definition of the term, *agricultural monoculture*, which is very similar to this answer.

48. C: This is the correct answer choice because there is ample evidence in the passage that refers to Carver's brilliance and the fact that his discoveries had a far-reaching impact both then and now. There is no evidence in the passage to support any of the other answer choices.

49. A: It introduces certain insects that transition from water to air. Choice *B* is incorrect because although the passage talks about gills, it is not the central idea of the passage. Choices *C* and *D* are incorrect because the passage does not "define" or "invite," but only serves as an introduction to stoneflies, dragonflies, and mayflies and their transition from water to air. Choice *E* is incorrect; the passage mentions "boatmen" and "scorpions" but does not explain their life expectancy.

50. C: The act of shedding part or all of the outer shell. Choices *A*, *B*, *D,* and *E* are incorrect.

Mathematics

1. B: First, subtract 4 from each side. This yields $6t = 12$. Now, divide both sides by 6 to obtain $t = 2$.

2. B: To be directly proportional means that $y = mx$. If *x* is changed from 5 to 20, the value of *x* is multiplied by 4. Applying the same rule to the y-value, also multiply the value of *y* by 4. Therefore, *y* = 12.

3. E: From the slope-intercept form, $y = mx + b$, it is known that *b* is the *y*-intercept, which is 1. Compute the slope as $\frac{2-1}{1-0} = 1$, so the equation should be $y = x + 1$.

4. A: Each bag contributes 4*x* + 1 treats. The total treats will be in the form $4nx + n$ where *n* is the total number of bags. The total is in the form 60*x* + 15, from which it is known $n = 15$.

5. D: Let *a* be the number of apples and *b* the number of bananas. Then, the total cost is $2a + 3b = 22$, while it also known that $a + b = 10$. Using the knowledge of systems of equations, cancel the *b* variables by multiplying the second equation by -3. This makes the equation $-3a - 3b = -30$. Adding this to the first equation, the b values cancel to get $-a = -8$, which simplifies to *a* = 8.

6. A: Finding the roots means finding the values of *x* when *y* is zero. The quadratic formula could be used, but in this case it is possible to factor by hand, since the numbers -1 and 2 add to 1 and multiply to -2. So, factor $x^2 + x - 2 = (x - 1)(x + 2) = 0$, then set each factor equal to zero. Solving for each value gives the values *x* = 1 and *x* = -2.

7. C: To find the *y*-intercept, substitute zero for *x*, which gives us:

$$y = 0^{\frac{5}{3}} + (0 - 3)(0 + 1)$$

$$0 + (-3)(1) = -3$$

8. E: This has the form $t^2 - y^2$, with $t = x^2$ and $y = 4$. It's also known that $t^2 - y^2 = (t + y)(t - y)$, and substituting the values for *t* and *y* into the right-hand side gives $(x^2 - 4)(x^2 + 4)$.

9. A: Simplify this to:

$$(4x^2y^4)^{\frac{3}{2}} = 4^{\frac{3}{2}}(x^2)^{\frac{3}{2}}(y^4)^{\frac{3}{2}}$$

Now:

$$4^{\frac{3}{2}} = (\sqrt{4})^3 = 2^3 = 8$$

For the other, recall that the exponents must be multiplied, so this yields:

$$8x^{2\cdot\frac{3}{2}}y^{4\cdot\frac{3}{2}} = 8x^3y^6$$

10. B: Start by squaring both sides to get $1 + x = 16$. Then subtract 1 from both sides to get $x = 15$.

11. C: Multiply both sides by *x* to get $x + 2 = 2x$, which simplifies to $-x = -2$, or *x* = 2.

12. B: The independent variable's coordinate at the vertex of a parabola (which is the highest point, when the coefficient of the squared independent variable is negative) is given by $x = -\frac{b}{2a}$. Substitute and solve for *x* to get:

$$x = -\frac{4}{2(-16)} = \frac{1}{8}$$

Using this value of *x*, the maximum height of the ball (*y*), can be calculated. Substituting *x* into the equation yields:

$$h(t) = -16\frac{1}{8}^2 + 4\frac{1}{8} + 6 = 6.25$$

13. D: Denote the width as *w* and the length as *l*. Then, $l = 3w + 5$. The perimeter is $2w + 2l = 90$. Substituting the first expression for *l* into the second equation yields:

$$2(3w + 5) + 2w = 90, \text{ or } 8w = 80, \text{ so} w l = 10$$

Putting this into the first equation, it yields:

$$l = 3(10) + 5 = 35$$

14. A: Lining up the given scores provides the following list: 60, 75, 80, 85, and one unknown. Because the median needs to be 80, it means 80 must be the middle data point out of these five. Therefore, the unknown data point must be the fourth or fifth data point, meaning it must be greater than or equal to 80. The only answer that fails to meet this condition is 60.

15. B: If 60% of 50 workers are women, then there are 30 women working in the office. If half of them are wearing skirts, then that means 15 women wear skirts. Since none of the men wear skirts, this means there are 15 people wearing skirts.

16. A: Let the unknown score be *x*. The average will be:

$$\frac{5 \cdot 50 + 4 \cdot 70 + x}{10} = \frac{530 + x}{10} = 55$$

Multiply both sides by 10 to get $530 + x = 550$, or $x = 20$.

17. E: For manufacturing costs, there is a linear relationship between the cost to the company and the number produced, with a *y*-intercept given by the base cost of acquiring the means of production, and a slope given by the cost to produce one unit. In this case, that base cost is $50,000, while the cost per unit is $40. So:

$$y = 40x + 50{,}000$$

18. C: A die has an equal chance for each outcome. Since it has six sides, each outcome has a probability of $\frac{1}{6}$. The chance of a 1 or a 2 is therefore $\frac{1}{6} + \frac{1}{6} = \frac{1}{3}$.

19. A: The slope is given by:

$$m = \frac{y_2 - y_1}{x_2 - x_1} = \frac{0 - 4}{0 - (-3)} = -\frac{4}{3}$$

20. B: An equilateral triangle has three sides of equal length, so if the total perimeter is 18 feet, each side must be 6 feet long. A square with sides of 6 feet will have an area of $6^2 = 36$ square feet.

21. A: 3.6

Divide 3 by 5 to get 0.6 and add that to the whole number 3, to get 3.6. An alternative is to incorporate the whole number 3 earlier on by creating an improper fraction: $\frac{18}{5}$. Then dividing 18 by 5 to get 3.6.

22. B: 300 miles in 4 hours is $\frac{300}{4}$ = 75 miles per hour. In 1.5 hours, the car will go 1.5×75 miles, or 112.5 miles.

23. A: Mean. An outlier is a data value that is either far above or far below the majority of values in a sample set. The mean is the average of all the values in the set. In a small sample set, a very high or very low number could drastically change the average of the data points. Outliers will have no more of an effect on the median (the middle value when arranged from lowest to highest) than any other value above or below the median. If the same outlier does not repeat, outliers will have no effect on the mode (value that repeats most often).

24. C: Line graph. The scenario involves data consisting of two variables, month and stock value. Box plots display data consisting of values for one variable. Therefore, a box plot is not an appropriate choice. Both line plots and circle graphs are used to display frequencies within categorical data. Neither can be used for the given scenario. Line graphs display two numerical variables on a coordinate grid and show trends among the variables. Pie charts usually show proportional data.

25. D: $\frac{1}{12}$. The probability of picking the winner of the race is $\frac{1}{4}\left(\frac{number\ of\ favorable\ outcomes}{number\ of\ total\ outcomes}\right)$. Assuming the winner was picked on the first selection, three horses remain from which to choose the runner-up (these are dependent events). Therefore, the probability of picking the runner-up is $\frac{1}{3}$. To determine the probability of multiple events, the probability of each event is multiplied:

$$\frac{1}{4} \times \frac{1}{3} = \frac{1}{12}$$

26. B: This inequality can be seen with the use of a number line. $\frac{3}{7}$ is close to $\frac{1}{2}$. $\frac{5}{6}$ is close to 1, but less than 1. Therefore, $\frac{3}{7}$ is less than $\frac{5}{6}$.

27. D: 104,165

Set up the problem and add each column, starting on the far right (ones). Add, carrying anything over 9 into the next column to the left. Solve from right to left.

28. E: 6.630

Set up the problem, with the larger number on top and numbers lined up at the decimal. Add, carrying anything over 9 into the next column to the left. Solve from right to left.

29. B: 148.97

Set up the problem, with the larger number on top and numbers lined up at the decimal. Insert 0 in any blank spots to the right of the decimal as placeholders. Add, carrying anything over 9 into the next column to the left.

30. C: Each number in the sequence is adding one more than the difference between the previous two. For example, $10 - 6 = 4, 4 + 1 = 5$. Therefore, the next number after 10 is $10 + 5 = 15$. Going forward, $21 - 15 = 6, 6 + 1 = 7$. The next number is $21 + 7 = 28$. Therefore, the difference between numbers is the set of whole numbers starting at 2: 2, 3, 4, 5, 6, 7....

31. C: $\frac{4}{3}$

Set up the problem and find a common denominator for both fractions.

$$\frac{14}{33} + \frac{10}{11}$$

Multiply each fraction across by 1 to convert to a common denominator

$$\frac{14}{33} \times \frac{1}{1} + \frac{10}{11} \times \frac{3}{3}$$

Once over the same denominator, add across the top. The total is over the common denominator.

$$\frac{14 + 30}{33} = \frac{44}{33}$$

Reduce by dividing both numerator and denominator by 11.

$$\frac{44 \div 11}{33 \div 11} = \frac{4}{3}$$

32. B: 128

This question involves the percent formula.

$$\frac{32}{x} = \frac{25}{100}$$

We multiply the diagonal numbers, 32 and 100, to get 3,200. Dividing by the remaining number, 25, gives us 128.

The percent formula does not have to be used for a question like this. Since 25% is ¼ of 100, you know that 32 needs to be multiplied by 4, which yields 128.

33. C: 40*N* would be 4000% of *N*. It's possible to check that each of the others is actually 40% of *N*.

34. E: 110,833

Set up the problem, with the larger number on top. Begin subtracting with the far-right column (ones). Borrow 10 from the column to the left, when necessary.

35. B: This is a division problem because the original amount needs to be split up into equal amounts. The mixed number $12\frac{1}{2}$ should be converted to an improper fraction first.

$$12\frac{1}{2} = \frac{(12 * 2) + 1}{2} = \frac{23}{2}$$

Carey needs to determine how many times $\frac{23}{2}$ goes into 184. This is a division problem:

$$184 \div \frac{23}{2} = ?$$

The fraction can be flipped, and the problem turns into the multiplication:

$$184 \times \frac{2}{23} = \frac{368}{23}$$

This improper fraction can be simplified into 16 because $368 \div 23 = 16$. The answer is 16 lawn segments.

36. B: 648.77

Set up the problem, with the larger number on top and numbers lined up at the decimal. Insert 0 in any blank spots to the right of the decimal as placeholders. Begin subtracting with the far-right column. Borrow 10 from the column to the left, when necessary.

37. A: The additive and subtractive identity is 0. When added or subtracted to any number, 0 does not change the original number.

38. B: 100 cm is equal to 1 m. 1.3 divided by 100 is 0.013. Therefore, 1.3 cm is equal to 0.013 mm. Because 1 cm is equal to 10 mm, 1.3 cm is equal to 13 mm.

39. E: 725

Set up the division problem.

$$1.4\overline{)1015}$$

Move the decimal over one place to the right in both numbers.

$$14\overline{)10150}$$

14 does not go into 1 or 10 but does go into 101 so start there.

$$\begin{array}{r} 725 \\ 14\overline{)10150} \\ \underline{-98} \\ 35 \\ \underline{-28} \\ 70 \\ \underline{-70} \\ 0 \end{array}$$

The result is 725.

40. C: One apple/orange pair costs $3 total. Therefore, Jan bought 90/3 = 30 total pairs, and hence, she bought 30 oranges.

41. B: 2.48

Set up the problem, with the larger number on top. Multiply as if there are no decimal places. Add the answer rows together. Count the number of decimal places that were in the original numbers ($1 + 1 = 2$).

Place the decimal 2 places to the right for the final solution.

42. B: 99.35

Set up the problem, with the larger number on top. Multiply as if there are no decimal places. Add the answer rows together. Count the number of decimal places that were in the original numbers (2).

Place the decimal in that many spots from the right for the final solution.

43. B: Each hour on the clock represents 30 degrees. For example, 3:00 represents a right angle. Therefore, 5:00 represents 150 degrees.

44. D: $27\frac{7}{22}$

Set up the division problem.

$$44\overline{)1202}$$

44 does not go into 1 or 12 but will go into 120 so start there.

$$\begin{array}{r} 27 \\ 44\overline{)1202} \\ -88 \\ \hline 322 \\ -308 \\ \hline 14 \end{array}$$

The answer is $27\frac{14}{44}$.

Reduce the fraction for the final answer.

$$27\frac{7}{22}$$

45. C: The formula for the volume of a box with rectangular sides is the length times width times height, so:

$$5 \times 6 \times 3 = 90 \text{ cubic feet}$$

46. E: 270

Set up the division problem.

$$2.6\overline{)702}$$

Move the decimal over one place to the right in both numbers.

$$26\overline{)7020}$$

26 does not go into 7 but does go into 70 so start there.

$$\begin{array}{r} 270 \\ 26\overline{)7020} \\ -52 \\ \hline 182 \\ -182 \\ \hline 0 \end{array}$$

The result is 270

47. D: First, the train's journey in the real word is 3 x 50 = 150 miles. On the map, 1 inch corresponds to 10 miles, so there is 150/10 = 15 inches on the map.

48. B: The total trip time is 1 + 3.5 + 0.5 = 5 hours. The total time driving is 1 + 0.5 = 1.5 hours. So, the fraction of time spent driving is 1.5/5 or 3/10. To get the percentage, convert this to a fraction out of 100. The numerator and denominator are multiplied by 10, with a result of 30/100. The percentage is the numerator in a fraction out of 100, so 30%.

49. B: The formula for the volume of a cylinder is $\pi r^2 h$, where *r* is the radius and *h* is the height. The diameter is twice the radius, so these barrels have a radius of 1 foot. That means each barrel has a volume of $\pi \times 1^2 \times 3 = 3\pi$ cubic feet. Since there are three of them, the total is:

$$3 \times 3\pi = 9\pi \text{ cubic feet}$$

50. A: The tip is not taxed, so he pays 5% tax only on the \$10. 5% of \$10 is $0.05 \times 10 = \$0.50$. Add up $\$10 + \$2 + \$0.50$ to get \$12.50.

CBEST Practice Test #2

Reading

1. After reading *To Kill a Mockingbird*, Louise has been asked to write an expository piece that explores the life, significant achievements, and societal impact of Harper Lee, the book's author. Which of the following sources would yield the most information about the author?

a. A dictionary
b. A newspaper article about the author
c. A study guide for *To Kill a Mockingbird*
d. A biographical account
e. An analysis on the life's work of the author

2. Read the following words:

mixed
thrown
are
grown
beaten
jumped

Analyze the list and determine which word does not belong.

a. Mixed
b. Are
c. Grown
d. Jumped
e. Beaten

3. Read the following words:

formaldehyde
forward
foliage
follicle
format
fort

Which of the following words would not be found in a dictionary between the guide words focus and fortitude?

a. Formaldehyde
b. Forge
c. Format
d. Fort
e. Forward

4. Asbestos was an *insidious* yet popular product, mass produced for over one hundred years. Due to its durability and fire resistance, it was used in a wide range of products, such as houses, cars, and ships. As with tobacco, evidence was presented early on that asbestos was dangerous and had cumulative adverse effects, but production didn't decline until the 1970s.

A dictionary provides four different definitions for *insidious*. Based on the above passage, which definition would fit best?

a. Awaiting a chance to trap or ensnare
b. Damaging or deadly but attractive
c. Having incremental or gradual build-up of harmful effects
d. Causing catastrophic harm
e. Causing a sudden explosion

5. After applying for a job multiple times, Bob was finally granted an interview. During the interview, he fumbled several questions, mispronounced his potential supervisor's name, and forgot the name of the company. Soon it was clear he knew he wouldn't be working there. At the end of the interview, he stood up and, with *spurious* confidence, shook the interviewer's hand firmly.

Based on the context of the word *spurious*, a good substitution might be:

a. Extreme
b. Mild
c. Proud
d. Genuine
e. Fake

6. Follow the numbered instructions to transform the starting word into a different word.

1. Start with the word KAKISTOCRACY.
2. Remove the first K.
3. Change the Y to an I.
4. Move the last C to the right of the last I.
5. Add a T between the last A and the last I.
6. Change the K to R.

What is the new word?

a. Aristocrat
b. Artistic
c. Aristocratic
d. Artifact
e. Antidote

Question 7 is based upon the following table.

Name brand diaper #1	Name brand diaper #2	U-Save discount diaper	Online diaper
88 diapers for $29.99	30 diapers for $12.00	50 diapers for $13.00	100 diapers for $45.00
$4.00 in tax	$2.00 in tax	$3.00 in tax	No tax

7. Marge, who is a new parent, is trying to find the best deal on diapers. She has priced them from several different locations. Based on the chart above, which product offers the best deal?

a. Name brand diaper #1
b. Name brand diaper #2
c. U-Save discount diaper
d. Online diaper
e. They all come out the same

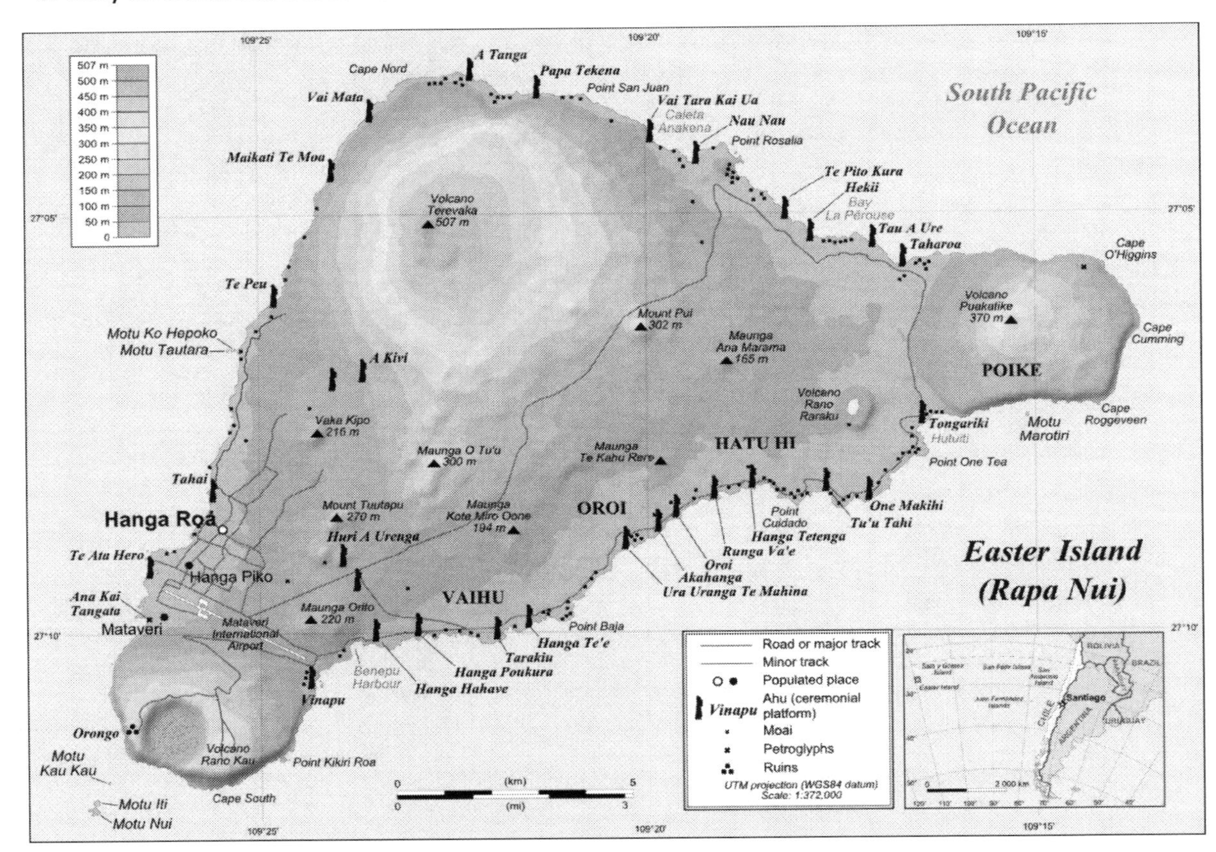

(Eric Gaba, *Wikimedia*, 2016)

Refer to the map above for questions 8–11.

8. How many land masses are represented on this map?

a. One
b. Two
c. Three
d. Four
e. Five

9. How many locations are at or above 300 meters?
   a. Four
   b. Five
   c. Six
   d. Seven
   e. Eight

10. Where is the most populated area of the island?
   a. In the northwest corner
   b. Centrally located
   c. In the southwest corner
   d. Near the coastline
   e. In the north central area

11. Where are most ruins on the island?
   a. In the northwest corner of the island
   b. Centrally located
   c. In the southwest corner of the island
   d. In the north central area
   e. Near the coastline

12. Reggie had been preparing for his part in the play for several months. His mother had even knitted him a costume. That night when he went on stage, his entire family and all his friends were there to watch, and he was *mortified* when his costume split in half. Afterwards, his mother admitted that she was not the best of seamstresses.

The best substitute for *mortified* would be which of the following?
   a. Inhibited
   b. Humiliated
   c. Annoyed
   d. Afraid
   e. Elated

Questions 13–16 are based on the following passage.

> Science fiction has been a part of the American fabric for a long time. During the Great Depression, many Americans were looking for an escape from dismal circumstances, and their escape often took the form of reading. Outlandish stories of aliens and superheroes were printed on cheap, disposable paper stock, hence the name *pulp* (as in paper) *fiction*. Iconic heroes like Buck Rogers, the Shadow, and Doc Rogers got their start in throwaway magazines and pulp novels.
>
> As time went on, science fiction evolved, presenting better plots and more sophisticated questions, and, consequently, it garnered more respect. Authors like Kurt Vonnegut and Ray Bradbury, now household names and respected American authors, emerged from the science fiction fringe. Thanks to works like Vonnegut's 1961 short story "Harrison Bergeron," in which mediocrity is the law and exceptional ability is punished, and Bradbury's 1953 novel *Fahrenheit 451*, in which books are illegal and burned on sight, science fiction rose to a serious genre.

In the late 1970s, the genre that begun in the medium of pulp fiction and crossed into serious literature had a resurgence in the medium of film. The new prominence of science fiction film was spearheaded by the first *Star Wars* movie, which harkened to pulp fiction roots. The tide of science fiction films hasn't really slowed since. *Blade Runner, Jurassic Park, The Matrix, I Am Legend*, even Disney's *Wall-E* all continue the tradition of unrealized futures and alternate realities. Modern science fiction movies can trace their roots back to the pulp fiction published during the Great Depression.

13. The main purpose of this passage is to:
 a. Describe
 b. Inform
 c. Persuade
 d. Entertain
 e. Instruct

14. This passage was written with a __________ structure.
 a. Compare/contrast
 b. Sequential
 c. Cause/effect
 d. Problem/solution
 e. Descriptive

15. Which of the following passages from the above text best summarizes the main idea?
 a. "Science fiction has been a part of the American fabric for a long time."
 b. "As time went on, science fiction evolved, posing better plots and more sophisticated questions."
 c. "Outlandish stories of aliens and superheroes were printed on cheap, disposable paper stock, hence the name *pulp* (as in paper) *fiction*."
 d. "Modern science fiction movies can trace their roots back to the pulp fiction published during the Great Depression."
 e. "Thanks to works like Vonnegut's 1961 short story 'Harrison Bergeron,' in which mediocrity is the law and exceptional ability is punished, and Bradbury's 1953 novel *Fahrenheit 451*, in which books are illegal and burned on sight, science fiction rose to a serious genre."

16. "These outlandish stories of aliens and superheroes were printed on cheap, disposable paper stock, hence the name *pulp* (as in paper) *fiction*." The author most likely wrote this sentence because:
 a. He or she wants the reader to understand that science fiction was not always taken seriously.
 b. He or she wants the reader to understand that science fiction is not a new medium.
 c. He or she wants to demonstrate that science fiction was constrained by the technology of the time.
 d. He or she wants to demonstrate that even those in the past imagined what the future might hold.
 e. He or she wants the reader to understand the importance that pulp fiction carried during its time.

Use the following definition to answer questions 17–19.

A dictionary provides the following information for the word *involve*: in volve (in-volv') v.t. [INVOLVED (-volvd'), INVOLVING], [M.E. *enoulen*; Ofr. *Involver*; L. *involvere*; *in-*, in + *volvere*, to roll up]

17. What does v.t. indicate?
    a. The origin(s) of the word
    b. How to pronounce the word
    c. How to use the word
    d. How many syllables are present
    e. In what region the word was used

18. What does the hyphen in the word *in-volve* indicate?
    a. The origins of the word
    b. How to pronounce the word
    c. How to use the word
    d. How many syllables are present
    e. How many letters are present

19. What are the oldest origins for the word *involve*?
    a. Middle English
    b. Old French
    c. Old Germanic
    d. Old English
    e. Latin

Use the table below to answer questions 20–23.

	Car 1	**Car 2**	**Car 3**	**Car 4**
Distance Traveled	100 miles	50 miles	70 miles	200 miles
Gallons Needed	4.23 gallons needed	2.083 gallons needed	3.181 gallons needed	8 gallons needed
Size of Tank	12-gallon tank	13-gallon tank	16-gallon tank	14-gallon tank

20. Based on the information above, which car gets the best gas mileage?
    a. Car 1
    b. Car 2
    c. Car 3
    d. Car 4
    e. Cars 3 & 4

21. Which car gets the worst gas mileage?
    a. Car 1
    b. Car 2
    c. Car 3
    d. Car 4
    e. Cars 1 & 2

22. Which car has the greatest range on one tank of gas?
    a. Car 1
    b. Car 2
    c. Car 3
    d. Car 4
    e. Cars 2 & 3

23. Which car has the least range on one tank of gas?
    a. Car 1
    b. Car 2
    c. Car 3
    d. Car 4
    e. Cars 1 & 3

Questions 24–27 are based on the following passage.

> The more immediately after the commission of a crime a punishment is inflicted, the more just and useful it will be. It will be more just, because it spares the criminal the cruel and superfluous torment of uncertainty, which increases in proportion to the strength of his imagination and the sense of his weakness; and because the privation of liberty, being a punishment, ought to be inflicted before condemnation, but for as short a time as possible. Imprisonments, I say, being only the means of securing the person of the accused, until he be tried, condemned, or acquitted, ought not only to be of as short duration, but attended with as little severity as possible. The time should be determined by the necessary preparation for the trial, and the right of priority in the oldest prisoners. The confinement ought not to be closer than is requisite to prevent his flight, or his concealing the proofs of the crime; and the trial should be conducted with all possible expedition. Can there be a more cruel contrast than that between the indolence of a judge, and the painful anxiety of the accused; the comforts and pleasures of an insensible magistrate, and the filth and misery of the prisoner? In general, as I have before observed, *the degree of the punishment, and the consequences of a crime, ought to be so contrived, as to have the greatest possible effect on others, with the least possible pain to the delinquent*. If there be any society in which this is not a fundamental principle, it is an unlawful society; for mankind, by their union, originally intended to subject themselves to the least evils possible.
>
> An immediate punishment is more useful; because the smaller the interval of time between the punishment and the crime, the stronger and more lasting will be the association of the two ideas of *Crime* and *Punishment;* so that they may be considered, one as the cause, and the other as the unavoidable and necessary effect. It is demonstrated, that the association of ideas is the cement which unites the fabric of the human intellect; without which, pleasure and pain would be simple and ineffectual sensations. The vulgar, that is, all men who have no general ideas or universal principles, act in consequence of the most immediate and familiar associations; but the more remote and complex only present themselves to the minds of those who are passionately attached to a single object, or to those of greater understanding, who have acquired an habit of rapidly comparing together a number of objects, and of forming a conclusion; and the result, that is, the action in consequence, by these means, becomes less dangerous and uncertain.

It is, then, of the greatest importance, that the punishment should succeed the crime as immediately as possible, if we intend, that, in the rude minds of the multitude, the seducing picture of the advantage arising from the crime, should instantly awake the attendant idea of punishment. Delaying the punishment serves only to separate these two ideas; and thus affects the minds of the spectators rather as being a terrible sight than the necessary consequence of a crime; the horror of which should contribute to heighten the idea of the punishment.

There is another excellent method of strengthening this important connexion between the ideas of crime and punishment; that is, to make the punishment as analogous as possible to the nature of the crime; in order that the punishment may lead the mind to consider the crime in a different point of view, from that in which it was placed by the flattering idea of promised advantages.

Crimes of less importance are commonly punished, either in the obscurity of a prison, or the criminal is *transported*, to give, by his slavery, an example to societies which he never offended; an example absolutely useless, because distant from the place where the crime was committed. Men do not, in general, commit great crimes deliberately, but rather in a sudden gust of passion; and they commonly look on the punishment due to a great crime as remote and improbable. The public punishment, therefore, of small crimes will make a greater impression, and, by deterring men from the smaller, will effectually prevent the greater.

(Cesare Beccaria, "Punishments, Advantages of Immediate," *The Criminal Recorder,* 1810).

24. What is the main purpose of this passage?
    a. To describe
    b. To inform
    c. To persuade
    d. To entertain
    e. To instruct

25. What text structure is this passage using?
    a. Compare/contrast
    b. Sequential
    c. Cause/effect
    d. Problem-solution
    e. Descriptive

26. Which of the following excerpts best exemplifies the main idea of this passage?
    a. "The vulgar, that is, all men who have no general ideas or universal principles, act in consequence of the most immediate and familiar associations."
    b. "Crimes of less importance are commonly punished, either in the obscurity of a prison, or the criminal is *transported*, to give, by his slavery, an example to societies which he never offended."
    c. "Men do not, in general, commit great crimes deliberately, but rather in a sudden gust of passion; and they commonly look on the punishment due to a great crime as remote and improbable."
    d. "The more immediately after the commission of a crime a punishment is inflicted, the more just and useful it will be."
    e. "The public punishment, therefore, of small crimes will make a greater impression, and, by deterring men from the smaller, will effectually prevent the greater."

27. With which of the following statements would the author most likely disagree?
    a. Criminals are incapable of connecting crime and punishment.
    b. A punishment should quickly follow the crime.
    c. Most criminals do not think about the consequences of their actions.
    d. Where a criminal is punished is just as important as when.
    e. A criminal should not be published in public.

Questions 28–30 are based on the following table.

Cellular Phone 1	**Cellular Phone 2**	**Cellular Phone 3**	**Cellular Phone 4**
Cost of phone: $200	Cost of phone: $100	Cost of phone: $50	Cost of phone: free
Monthly plan: $35	Monthly plan: $25	Monthly plan: $50	Monthly plan: $65
Monthly insurance: $20	Monthly insurance: $30	Monthly insurance: $25	Monthly insurance: $7

28. Taking into consideration the cost of the phone and payments per month, which contract would offer the best value over the course of two months?
    a. Cellular phone 1
    b. Cellular phone 2
    c. Cellular phone 3
    d. Cellular phone 4
    e. Both cellular phones 1 & 2

29. Taking into consideration the cost of the phone and payments per month, which contract would offer the best value over the course of a year?
    a. Cellular phone 1
    b. Cellular phone 2
    c. Cellular phone 3
    d. Cellular phone 4
    e. Both cellular phones 3 & 4

30. If someone were accident prone and needed the monthly insurance, which phone would offer the best value over the course of a year?
    a. Cellular phone 1
    b. Cellular phone 2
    c. Cellular phone 3
    d. Cellular phone 4
    e. Both cellular phones 1 & 3

Questions 31–32 are based on the following passage.

Dear Brand-X Employees:

Over the past ten years, Brand-X has been happy to provide free daycare to Brand-X employees. Based on the dedication, long workdays, and professionalism shown daily, it's the least Brand-X management can do. Brand-X wouldn't be where it is today without the hard work of countless, unsung heroes and is truly blessed to have such an amazing staff.

Unfortunately, Brand-X is subject to the same economic forces as any other company. We regret to inform you that, beginning March 15th, Brand-X has decided to discontinue the childcare program. This was a difficult decision to make, one that was arrived at only with the greatest of deliberation. Many other options were discussed, and this seemed to be the only one that allowed Brand-X to still stay competitive while not reducing staff.

Fortunately, all other programs—employee rewards, vacation and sick days, retirement options, volunteer days—have not been impacted. In addition, on Friday we'll be hosting a free lunch buffet. Employees are welcome to wear jeans.

Hope to see you there,

Brand-X Management

31. Which of the following words best describe the tone in the first, second, and third paragraphs?
    a. Relieved; conciliatory; ominous
    b. Conciliatory; ominous; relieved
    c. Proud; apologetic; placating
    d. Placating; apologetic; proud
    e. Apologetic; placating; relieved

32. "Unfortunately, Brand-X is subject to the same economic forces as any other company." This line was most likely provided to do what?
    a. Emphasize that Brand-X is no worse than other companies.
    b. Redirect blame from Brand-X to outside forces.
    c. Encourage staff to look for work elsewhere.
    d. Intimidate staff into cooperating.
    e. Insult Brand-X to make the employees feel better.

For his American literature course next fall, Benito's professor submits a required summer reading list to all students. Before the first day of class, Benito is required to read *The Sound and the Fury*, "Barn Burning," *For Whom the Bell Tolls*, *The Grapes of Wrath*, and "A&P."

33. How many books is Benito required to read during the summer?
    a. One
    b. Two
    c. Three
    d. Four
    e. Five

34. How many short stories is Benito required to read during the summer?
    a. One
    b. Two
    c. Three
    d. Four
    e. Five

35. Carl's chore for the day was to clean out a little-used storage shed. When he unlocked the door and walked through it, he got a mouthful of spider web. Brushing his mouth and sputtering, he reached for the light's pull chain. As soon as the bulb went on, a mouse ran between his legs and out the door. *This place is worse than a zoo*, thought Carl as he reached *reluctantly* for a box.

The best replacement for *reluctantly* would be which of the following?

a. Angrily
b. Suspiciously
c. Fearfully
d. Excitingly
e. Hesitantly

36. Jason had never been the best student. He had flunked three math tests and turned in only half of his homework. Worrying about his graduation, Jason's mother signed him up for tutoring. Even after several sessions, it was unclear whether he would pass. In the last week of math class, Jason buckled down and passed. Jason's mother marveled at his *gumption*.

The best substitute for *gumption* would be which of the following?

a. Stubbornness
b. Laziness
c. Initiative
d. Ambition
e. Courage

Questions 37 and 38 are based on the following table.

Ship 1	**Ship 2**	**Ship 3**	**Ship 4**
Depart: 1:10 p.m.	Depart: 1:00 p.m.	Depart: 1:30 p.m.	Depart: 12:30 p.m.
Arrive: 2:30 p.m.	Arrive: 2:45 p.m.	Arrive: 2:20 p.m.	Arrive: 1:50 p.m.
Return: 4:40 p.m.	Return: 5:30 p.m.	Return: 5:30 p.m.	Return: 5:45 p.m.

37. Lucy and Bob both enjoy fishing and want to take a charter ship to an island, but they have different schedules. Lucy, who works mornings, can't leave until 12:45 p.m. She needs thirty minutes to arrive at the dock. Bob, on the other hand, starts work at 6:30 p.m. and needs an hour to get from the docks to his job. There are four different charter ships available. Based on their schedules, which ship would meet both Lucy and Bob's needs?

a. Ship 1
b. Ship 2
c. Ship 3
d. Ship 4
e. Ship 1 & 2

38. If Lucy and Bob didn't have any time restraints, which boat would give them the most time on the island to fish?

a. Ship 1
b. Ship 2
c. Ship 3
d. Ship 4
e. Ship 3 & 4

Questions 39–42 are based on the following two passages.

Passage 1

In the modern classroom, cell phones have become indispensable. Cell phones, which are essentially handheld computers, allow students to take notes, connect to the web, perform complex computations, teleconference, and participate in surveys. Most importantly, though, due to their mobility and excellent reception, cell phones are necessary in emergencies. Unlike tablets, laptops, or computers, cell phones are a readily available and free resource—most school district budgets are already strained to begin with—and since today's student is already strongly rooted in technology, when teachers incorporate cell phones, they're "speaking" the student's language, which increases the chance of higher engagement.

Passage 2

As with most forms of technology, there is an appropriate time and place for the use of cell phones. Students are comfortable with cell phones, so it makes sense when teachers allow cell phone use at their discretion. Allowing cell phone use can prove advantageous if done correctly. Unfortunately, if that's not the case—and often it isn't—then a sizable percentage of students pretend to pay attention while *surreptitiously* playing on their phones. This type of disrespectful behavior is often justified by the argument that cell phones are not only a privilege but also a right. Under this logic, confiscating phones is akin to rummaging through students' backpacks. This is in stark contrast to several decades ago when teachers regulated where and when students accessed information.

39. With which of the following statements would both the authors of Passages 1 and 2 agree?
   a. Teachers should incorporate cell phones into curriculum whenever possible.
   b. Cell phones are useful only when an experienced teacher uses them properly.
   c. Cell phones and, moreover, technology, are a strong part of today's culture.
   d. Despite a good lesson plan, cell phone disruptions are impossible to avoid.
   e. Cell phones are necessary in an emergency.

40. Which of the following reasons is NOT listed in Passage 1 as a reason for students to have cell phones?
   a. Cell phones are a free, readily available resource.
   b. Cell phones incorporate others forms of technology.
   c. Due to their mobility, cell phones are excellent in an emergency.
   d. Cell phones allow teachers to "speak" clearly with students.
   e. Cell phones allow higher engagement between students and teachers.

41. Passage 2 includes the statement, "confiscating phones is akin to rummaging through students' backpacks." The author most likely included this statement to do which of the following?
   a. Indicate how unlikely students are to change their minds.
   b. Exemplify how strongly students believe this is a right.
   c. Exemplify how easily the modern student is offended.
   d. Demonstrate how illogical most students' beliefs are.
   e. Explain how the rules of education are rapidly changing.

42. Based on the context of Passage 2, the best substitute for *surreptitiously* would most likely be which of the following?

a. Privately
b. Casually
c. Obstinately
d. Defiantly
e. Slyly

Questions 43 and 44 are based on the following table.

Cooking Oils	Smoking Point F°	Neutral Taste?
Clarified Butter	485°	No
Peanut Oil	450°	Yes
Lard	374°	No
Safflower Oil	510°	Yes
Coconut Oil	350°	No

43. Zack is getting ready to heat some cooking oil. He knows that if an oil goes above its smoking point, it doesn't taste good. For his recipe, he must get the oil to reach 430° F. Zack has a peanut allergy and would prefer a neutral-tasting oil. Which oil should he use?

a. Clarified butter
b. Peanut oil
c. Safflower oil
d. Coconut oil
e. Lard

44. If Zack still needed the oil to reach 430° F but he didn't have a peanut allergy and preferred a flavored oil, which oil would he use?

a. Clarified butter
b. Peanut oil
c. Safflower oil
d. Coconut oil
e. Lard

45. After many failed attempts, Julio made a solemn promise to his mother to clean his room. When she came home from a long day of work, she found her son playing video games, his room still a disaster. Her fists clenched, her eyebrows knitted, she stared down her son and delivered a *vituperative* speech that made him hang his head regretfully.

The best substitute for *vituperative* would be which of the following?

a. Annoyed
b. Abusive
c. Passionate
d. Sorrowful
e. Virtuous

Questions 46–48 are based on the following passage.

Random Lake Advertisement

Who needs the hassle of traveling far away? This summer, why not rent a house on Random Lake? Located conveniently ten miles away from Random City, Random Lake has everything a family needs. Swimming, kayaking, boating, fishing, volleyball, mini-golf, go-cart track, eagle watching, nature trails—there are enough activities here to keep a family busy for a month, much less a week.

Random Lake hotels are available for every lifestyle and budget. Prefer a pool or free breakfast? Prefer quiet? Historical? Modern? No problem. The Random Lake area has got you covered. House rentals are affordable too. During the summer months, rentals can go as cheaply as 600 dollars a week. Even better deals can be found during the off season. Most homes come fully furnished, and pontoon boats, kayaks, and paddle boats are available for rental. With the Legends and the Broadmoor developments slated for grand openings in March, the choices are endless!

46. The main purpose of this passage is to do what?
   a. Describe
   b. Inform
   c. Narrate
   d. Entertain
   e. Persuade

47. Which of the following sentences is out of place and should be removed?
   a. "This summer, why not rent a house on Random Lake?"
   b. "There are enough activities here to keep a family busy for a month."
   c. "Pontoon boats, kayaks, and paddle boats are available for rental."
   d. "With several new housing developments slated for grand opening in March, the choices are endless!"
   e. "With the Legends and the Broadmoor developments slated for grand openings in March, the choices are endless!"

48. Which of the following can be deduced from Passages 1 and 2?
   a. Random City is more populated than Random Lake.
   b. The Random Lake area is newer than Random City.
   c. The Random Lake area is growing.
   d. Random Lake prefers families to couples.
   e. You can go kayaking at a Random Lake Hotel.

Questions 49–50 are based upon the following passage:

This excerpt is adaptation from "What to the Slave is the Fourth of July?" Rochester, New York, July 5, 1852.

Fellow citizens—Pardon me, and allow me to ask, why am I called upon to speak here today? What have I, or those I represent, to do with your national independence? Are the great principles of political freedom and of natural justice embodied in that Declaration of Independence, Independence extended to us? And am I therefore called upon to bring our humble offering to the national altar, and to confess the benefits, and express devout gratitude for the blessings, resulting from your independence to us?

Would to God, both for your sakes and ours, ours that an affirmative answer could be truthfully returned to these questions! Then would my task be light, and my burden easy and delightful. For who is there so cold that a nation's sympathy could not warm him? Who so obdurate and dead to the claims of gratitude that would not thankfully acknowledge such priceless benefits? Who so stolid and selfish, that would not give his voice to swell the hallelujahs of a nation's jubilee, when the chains of servitude had been torn from his limbs? I am not that man. In a case like that, the dumb may eloquently speak, and the lame man leap as an hart.

But, such is not the state of the case. I say it with a sad sense of the disparity between us. I am not included within the pale of this glorious anniversary. Oh pity! Your high independence only reveals the immeasurable distance between us. The blessings in which you this day rejoice, I do not enjoy in common. The rich inheritance of justice, liberty, prosperity, and independence, bequeathed by your fathers, is shared by *you*, not by *me*. This Fourth of July is *yours,* not *mine*. You may rejoice, *I* must mourn. To drag a man in fetters into the grand illuminated temple of liberty, and call upon him to join you in joyous anthems, were inhuman mockery and sacrilegious irony. Do you mean, citizens, to mock me, by asking me to speak today? If so there is a parallel to your conduct. And let me warn you that it is dangerous to copy the example of a nation whose crimes, towering up to heaven, were thrown down by the breath of the Almighty, burying that nation and irrecoverable ruin! I can today take up the plaintive lament of a peeled and woe-smitten people.

By the rivers of Babylon, there we sat down. Yea! We wept when we remembered Zion. We hanged our harps upon the willows in the midst thereof. For there, they that carried us away captive, required of us a song; and they who wasted us required of us mirth, saying, “Sing us one of the songs of Zion.” How can we sing the Lord's song in a strange land? If I forget thee, O Jerusalem, let my right hand forget her cunning. If I do not remember thee, let my tongue cleave to the roof of my mouth.

49. What is the tone of the first paragraph of this passage?
   a. Exasperated
   b. Inclusive
   c. Contemplative
   d. Nonchalant
   e. Irate

50. Which word CANNOT be used synonymously with the term *obdurate* as it is conveyed in the text below?

> Who so obdurate and dead to the claims of gratitude, that would not thankfully acknowledge such priceless benefits?

a. Steadfast
b. Stubborn
c. Adamant
d. Unwavering
e. Contented

Mathematics

1. Add $5{,}089 + 10{,}323$
 a. 15,402
 b. 15,412
 c. 5,234
 d. 15,234
 e. 10,534

2. A teacher is showing students how to evaluate $5 \times 6 + 4 \div 2 - 1$. Which operation should be completed first?
 a. Multiplication
 b. Addition
 c. Division
 d. Subtraction
 e. Exponentiation

3. What is the definition of a factor of the number 36?
 a. A number that can be divided by 36 and have no remainder
 b. A number that can be added to 36 with no remainder
 c. A prime number that is multiplied times 36
 d. An even number that is multiplied times 36
 e. A number that 36 can be divided by and have no remainder

4. Which of the following is the definition of a prime number?
 a. A number that factors only into itself and 1
 b. A number greater than zero that factors only into itself and 1
 c. A number less than 10
 d. A number divisible by 10
 e. A number divisible by 0

5. Add and express in reduced form $\frac{5}{12} + \frac{4}{9}$

a. $\frac{9}{17}$

b. $\frac{1}{3}$

c. $\frac{31}{36}$

d. $\frac{3}{5}$

e. $\frac{1}{5}$

6. Which of the following is the correct order of operations that could be used on a difficult math problem that contained grouping symbols?

a. Parentheses, Exponents, Multiplication, Division, Addition, Subtraction
b. Exponents, Parentheses, Multiplication, Division, Addition, Subtraction
c. Parentheses, Exponents, Addition, Multiplication, Division, Subtraction
d. Parentheses, Exponents, Division, Addition, Subtraction, Multiplication
e. Division, Exponents, Addition, Parentheses, Subtraction, Multiplication

7. Convert $\frac{5}{8}$ to a decimal.

a. 0.62
b. 1.05
c. 0.63
d. 1.60
e. 2.38

8. Subtract $9{,}576 - 891$.

a. 10,467
b. 9,685
c. 8,325
d. 7,833
e. 8,685

9. If a teacher was showing a class how to round 245.2678 to the nearest thousandth, which place value would be used to decide whether to round up or round down?

a. Ten-thousandth
b. Thousandth
c. Hundredth
d. Thousand
e. Tenth

10. Subtract $50.888 - 13.091$.

a. 63.799
b. 63.979
c. 37.979
d. 33,817
e. 37.797

11. Students should line up decimal places within the given numbers before performing which of the following?

a. Multiplication
b. Division
c. Subtraction
d. Exponents
e. Addition

12. Subtract and express in reduced form $\frac{23}{24} - \frac{1}{6}$.

a. $\frac{22}{18}$

b. $\frac{11}{9}$

c. $\frac{19}{24}$

d. $\frac{4}{5}$

e. $\frac{5}{6}$

13. Subtract and express in reduced form $\frac{43}{45} - \frac{11}{15}$.

a. $\frac{10}{45}$

b. $\frac{16}{15}$

c. $\frac{32}{30}$

d. $\frac{2}{9}$

e. $\frac{1}{3}$

14. Change 0.56 to a fraction.

a. $\frac{5.6}{100}$

b. $\frac{14}{25}$

c. $\frac{56}{1000}$

d. $\frac{56}{10}$

15. Multiply $13,114 \times 191$.

a. 2,504,774
b. 250,477
c. 150,474
d. 2,514,774
e. 2,515,774

16. Marty wishes to save $150 over a 4-day period. How much must Marty save each day on average?
    a. $37.50
    b. $35
    c. $45.50
    d. $41
    e. $42

17. A teacher cuts a pie into 6 equal pieces and takes one away. What topic would she be introducing to the class by using such a visual?
    a. Decimals
    b. Addition
    c. Subtraction
    d. Measurement
    e. Fractions

18. Multiply and reduce $\frac{15}{23} \times \frac{54}{127}$.
    a. $\frac{810}{2,921}$
    b. $\frac{81}{292}$
    c. $\frac{69}{150}$
    d. $\frac{810}{2929}$
    e. $\frac{30}{882}$

19. Which of the following represent one hundred eighty-two billion, thirty-six thousand, four hundred twenty-one and three hundred fifty-six thousandths?
    a. 182,036,421.356
    b. 182,036,421.0356
    c. 182,000,036,421.0356
    d. 182,000,036,421.356
    e. 182,036,000,421.356

20. Divide, express with a remainder $188 \div 16$.
    a. $1\frac{3}{4}$
    b. $111\frac{3}{4}$
    c. $10\frac{3}{4}$
    d. $11\frac{3}{4}$
    e. $3\frac{1}{111}$

21. What other operation could be utilized to teach the process of dividing 9453 by 24 besides division?
   a. Multiplication
   b. Addition
   c. Exponents
   d. Subtraction
   e. Parentheses

22. Bernard can make $80 per day. If he needs to make $300 and only works full days, how many days will this take?
   a. 6
   b. 3
   c. 5
   d. 4
   e. 2

23. A couple buys a house for $150,000. They sell it for $165,000. By what percentage did the house's value increase?
   a. 18%
   b. 13%
   c. 15%
   d. 17%
   e. 10%

24. What operation are students taught to repeat to evaluate an expression involving an exponent?
   a. Addition
   b. Multiplication
   c. Division
   d. Subtraction
   e. Parentheses

25. Which of the following formulas would correctly calculate the perimeter of a legal-sized piece of paper that is 14 inches long and $8\frac{1}{2}$ inches wide?
   a. $P = 14 + 8\frac{1}{2}$
   b. $P = 14 + 8\frac{1}{2} + 14 + 8\frac{1}{2}$
   c. $P = 14 \times 8\frac{1}{2}$
   d. $P = 14 \times \frac{17}{2}$
   e. $P = 8 \times 14 + \frac{1}{2}$

26. Which of the following are units that would be taught in a lecture covering the metric system?
   a. Inches, feet, miles, pounds
   b. Millimeters, centimeters, meters, pounds
   c. Kilograms, grams, kilometers, meters
   d. Teaspoons, tablespoons, ounces
   e. Minutes, seconds, milliseconds

27. Which important mathematical property is shown in the expression: $(7 \times 3) \times 2 = 7 \times (3 \times 2)$?
   a. Distributive property
   b. Commutative property
   c. Additive inverse
   d. Multiplicative inverse
   e. Associative property

28. A grocery store is selling individual bottles of water, and each bottle contains 750 milliliters of water. If 12 bottles are purchased, what conversion will correctly determine how many liters that customer will take home?
   a. 100 milliliters equals 1 liter
   b. 1,000 milliliters equals 1 liter
   c. 1,000 liters equals 1 milliliter
   d. 10 liters equals 1 milliliter
   e. 1 liter equals 2000 milliliters

29. If a student evaluated the expression $(3 + 7) - 6 \div 2$ to equal 2 on an exam, what error did she most likely make?
   a. She performed the operations from left to right instead of following order of operations.
   b. There was no error. 2 is the correct answer.
   c. She did not perform the operation within the grouping symbol first.
   d. She divided first instead of the addition within the grouping symbol.
   e. She performed the operations from right to left instead of following order of operations.

30. What is the solution to $(2 \times 20) \div (7 + 1) + (6 \times 0.01) + (4 \times 0.001)$?
   a. 5.064
   b. 5.64
   c. 5.0064
   d. 48.064
   e. 52.587

31. A cereal box has a base 3 inches by 5 inches and is 10 inches tall. Another box has a base 5 inches by 6 inches. What formula is necessary for students to use to find out how tall the second box would need to be in order to hold the same amount of cereal?
   a. Area of a rectangle
   b. Area of a circle
   c. Volume of a cube
   d. Perimeter of a square
   e. Volume of a rectangular solid

32. An angle measures 54 degrees. In order to correctly determine the measure of its complementary angle, what concept is necessary?
   a. Two complementary angles sum up to 180 degrees.
   b. Complementary angles are always acute.
   c. Two complementary angles sum up to 90 degrees.
   d. Complementary angles sum up to 360 degrees.
   e. Two complementary angles sum up to 180 degrees.

33. Which is closest to 17.8×9.9?
   a. 140
   b. 180
   c. 200
   d. 350
   e. 400

34. A school has 15 teachers and 20 teaching assistants. They have 200 students. What is the ratio of faculty to students?
   a. 3:20
   b. 4:17
   c. 5:54
   d. 7:40
   e. 8:40

35. $\frac{3}{4}$ of a pizza remains on the stove. Katie eats $\frac{1}{3}$ of the remaining pizza. In order to determine how much of the pizza is left, what topic must be introduced to the students?
   a. Converting fractions to decimals
   b. Subtraction of fractions with like denominators
   c. Addition of fractions with unlike denominators
   d. Division of fractions
   e. Adding exponents

36. Taylor works two jobs. The first pays $20,000 per year. The second pays $10,000 per year. She donates 15% of her income to charity. How much does she donate each year?
   a. $4500
   b. $5000
   c. $5500
   d. $6000
   e. $6500

37. Joshua has collected 12,345 nickels over a span of 8 years. He took them to bank to deposit into his bank account. If the students were asked to determine how much money he deposited, for what mathematical topic would this problem be a good introduction?
   a. Adding decimals
   b. Multiplying decimals
   c. Geometry
   d. The metric system
   e. Fractions

38. A box with rectangular sides is 24 inches wide, 18 inches deep, and 12 inches high. What is the volume of the box in cubic feet?
   a. 2
   b. 6
   c. 4
   d. 5
   e. 3

39. What is the solution to $9 \times 9 \div 9 + 9 - 9 \div 9$?
    a. 0
    b. 13
    c. 81
    d. 9
    e. 17

40. A student answers a problem with the following fraction: $\frac{3}{15}$. Why would this be considered incorrect?
    a. It is not expressed in decimal form.
    b. It is not simplified. The correct answer would be $\frac{1}{5}$.
    c. It needs to be converted to a mixed number.
    d. It is in the correct form, and there is no problem with it.
    e. It does not have an exponent.

41. The hospital has a nurse to patient ratio of 1:25. If there are a maximum of 325 patients admitted at a time, how many nurses are there?
    a. 13 nurses
    b. 25 nurses
    c. 325 nurses
    d. 12 nurses
    e. 50 nurses

42. A hospital has a bed to room ratio of $2:1$. If there are 145 rooms, how many beds are there?
    a. 145 beds
    b. 2 beds
    c. 90 beds
    d. 290 beds
    e. 300 beds

43. Solve for x: $\frac{2x}{5} - 1 = 59$.
    a. 60
    b. 145
    c. 150
    d. 115
    e. 130

44. A National Hockey League store in the state of Michigan advertises 50% off all items. Sales tax in Michigan is 6%. How much would a hat originally priced at $32.99 and a jersey originally priced at $64.99 cost during this sale? Round to the nearest penny.
    a. $97.98
    b. $103.86
    c. $51.93
    d. $48.99
    e. $44.98

45. Store brand coffee beans cost $1.23 per pound. A local coffee bean roaster charges $1.98 per 1 ½ pounds. How much more would 5 pounds from the local roaster cost than 5 pounds of the store brand?
a. $0.55
b. $1.55
c. $1.45
d. $0.45
e. $2.15

46. Paint Inc. charges $2000 for painting the first 1,800 feet of trim on a house and $1.00 per foot for each foot after. How much would it cost to paint a house with 3125 feet of trim?
a. $3125
b. $2000
c. $5125
d. $3325
e. $3500

47. A bucket can hold 11.4 liters of water. A kiddie pool needs 35 gallons of water to be full. How many times will the bucket need to be filled to fill the kiddie pool?
a. 12
b. 35
c. 11
d. 45
e. 50

48. Mom's car drove 72 miles in 90 minutes. There are 5280 feet per mile. How fast did she drive in feet per second?
a. 0.8 feet per second
b. 48.9 feet per second
c. 0.009 feet per second
d. 70. 4 feet per second
e. 9 feet per second

49. Convert 0.351 to a percentage.
a. 3.51%
b. 35.1%
c. $\frac{351}{100}$
d. 0.00351%
e. $\frac{1}{351}$

50. Convert $\frac{2}{9}$ to a percentage.
a. 22%
b. 4.5%
c. 450%
d. 0.22%
e. 78%

Writing

First Essay Prompt

Directions: The Writing section of the CBEST will require test takers to write two essays based on provided prompts. The first essay has a referential aim so that test takers can showcase their analytical and expository writing skills. The second essay will have an expressive aim, with a topic relating to the test taker's past lived experience.

Please read the prompt below and answer in an essay format.

> Margaret Atwood says "war is what happens when language fails." In an essay that is going to be read by educated adults, explain whether you agree or disagree with this observation. Support your argument with details and examples.

Second Essay Prompt

Please read the prompt below and answer in an essay format.

> Students and teachers alike have had mentors who have changed the course of their lives for the better. Name someone who was a mentor to you. Explain what they were like and how their experience and encouragement changed the course of your life.

Answer Explanations #2

Reading

1. D: To find the correct answer, ask what yields the most information and is relevant to the task at hand: finding out about Harper Lee. A dictionary's main purpose is to define words, provide tenses, and establish pronunciation. A newspaper article might offer some information about Harper Lee, but it would be limited in scope to a specific topic or time period. A study guide would focus on literary elements of *To Kill a Mockingbird*, not Harper Lee's life. A biography would be the most comprehensive. It would cover the author, from birth to death, and touch on topics such as upbringing, significant achievements, and societal impacts. An analysis of the author's life's work would be about her work, not her life.

2. B: Unlike all the other words on the list, the word *are* is a being verb. In addition, *are* is in the present tense. The words *mixed*, *thrown*, *grown*, *beaten*, and *jumped* are all action verbs, and they are all in the past tense. The past tense of *are*, of course, is *were*. Therefore, *are* does not fit for two reasons: verb type and tense.

3. E: *Forward* is the correct answer. Since *focus* and *fortitude* are the guide words, all words that fall between them must correspond alphabetically. *Formaldehyde*, beginning with an *f, o*, and *r*, falls after focus, but the next letter, *a*, places it before *fortitude*. In *forge*, the *f* and *o* falls after focus, but then the *g* comes before the *t* in *fortitude.* The *f, o,* and *r* in *format* are similar to *formaldehyde*, and, again, the *m* places it before *fortitude*. Then there's *fort*. Since the first four letters match fortitude but there's no additional letters, it would precede fortitude as an entry.

4. C: "Having incremental or gradual build-up of harmful effects" is the correct answer. To find the correct answer, try replacing or substituting words. Here, one might try saying "Damaging or deadly but attractive"; this answer is incorrect because, although the paragraph says the product was popular, it does not indicate that *insidious* means *popular*. "Awaiting a chance to entrap or ensnare" is incorrect because asbestos is an inanimate object. "Causing catastrophic harm" doesn't work because the level of harm was not specified in the paragraph, and "Causing a sudden explosion" is incorrect because the passage says that it was fire resistant.

5. E: To arrive at the correct answer, find the best synonym for *spurious*. *Fake* makes the most sense because Bob is trying to hide his true emotions. Based on his horrible performance, he understands that he didn't get the job, so the reader can infer that his confidence is not at an all-time high. Therefore, *fake* would make just as much sense in this case as *spurious*. *Extreme* would indicate that Bob is euphoric and the interview went well. *Mild* would indicate that Bob thought he had a slight chance at getting hired. In this instance, *genuine* would be an antonym; i.e., the exact opposite of what Bob is feeling. *Proud* would also not fit here, because Bob's interview went poorly.

6. C:

1. Start with the word KAKISTOCRACY
2. Replace the first K with the first A (akistocracy)
3. Change the Y to an I. (akistocraci)
4. Move the last C to the right of the last I (akistocraic)
5. Add a T between the last A and the last I. (akistocratic)
6. Change the K to R. (aristocratic)

7. C: U-Save discount diaper offers the best deal. To arrive at the correct answer, take the price of the diaper in each column and divide by the number of diapers. Next, take the cost of tax in each column and divide by the numbers of diapers. Add these two totals together to arrive at the total cost per individual diaper. (Note: for the online diaper column, there is no tax to calculate.) The u-save diaper has a total cost of $.26 cents per diaper and $.06 cents tax, totaling $.32 cents per diaper. The first name brand diaper has a cost of $.34 cents per diaper with $.04 cents tax per diaper, totaling $.38 cents per diaper. The second name brand diaper has $.40 cents per diaper and $.06 cents tax per diaper, totaling $.46 cents total per diaper. The online diaper has a cost of $.45 cents per diaper.

8. D: There are four land masses depicted on the map: Easter Island, Motu Kau Kau, Motu Iti, and Motu Nui.

9. A: According to both the number markings and the colored topography, there are four areas above 300 meters: Volcano Terevaka, Volcano Puakatike, Maunga O Tu'u, and Volcano Rano Kau.

10. C: Taking into consideration both the number of roads and the population marker (o) at Hanga Roa, the southwest corner of the island is the most heavily populated.

11. E: Most ruins are located near the coastline, as indicated by the ruin markers on the key.

12. B: *Humiliated* would be the best substitute for *mortified*. *Inhibited* is a general characteristic of a person unable to act in a relaxed or natural manner; it might be true of Reggie in general but does not describe his acute emotional state at that moment described in the anecdote. *Annoyed* is also too mild a word. *Afraid* is not a synonym of *mortified*. *Elated* means happy so this is incorrect.

13. B: The main purpose of this passage is to inform. It wants the reader to understand how American science fiction evolved. Choice *A* is incorrect because the passage is not very descriptive; there is not an abundance of adjectives and adverbs painting a picture in the reader's mind. The passage is not persuasive (Choice *C*); the author is not asking the reader to adopt a stance or argument. For Choice *D*, it might be mildly entertaining, but the entertainment aspect of the passage takes a back seat to its informative aspects. Choice *E*, *instruct*, is something an instruction manual or a technical paper would do, so this isn't the *best* answer.

14. B: This passage begins with the oldest (the pulp fiction magazines of the Great Depression), moves forward to the science fiction of the fifties and sixties via names like Ray Bradbury and Kurt Vonnegut, and ends with a discussion of modern science fiction movies—*Jurassic Park, The Matrix, I am Legend*. It follows a sequential timeline of American science fiction.

15. A: All other concepts discussed throughout this passage connect to the main idea: "Science fiction has been a part of the American fabric for a long time." Choice *B* states: "As time went on, science fiction evolved, posing better plots and more sophisticated questions." This is only a portion of the entire history of American science fiction. The same is true of Choice *C*: "These outlandish stories of aliens and superheroes were printed on cheap, disposable paper stock, hence the name *pulp* (as in paper) *fiction*." Choice *D*, though it ties together the Great Depression and modern American cinema, doesn't touch on the period in between. Choice *E* presents some supporting details of the main idea.

16. A: There are several clues that support Choice *A*. First, the fact that the stories were printed on *disposable paper* proves that these stories were, at the time, not considered serious literature. Furthermore, it's noted later in the passage that science fiction *was taken seriously* as authors like Kurt Vonnegut and Ray Bradbury gained prominence. This further reinforces the notion that at the beginning it was considered only fanciful and for children. Choice *B* is incorrect. Even though the statement is true of the passage, it's not supported by the individual sentence. Choice *C* is incorrect because the sentence never states directly or hints at this, and, in addition, movies and radio were available during this time period. Choice *D* is incorrect. The premise of the whole passage is that the medium of science fiction is not new. It doesn't spend any time discussing *how* science fiction was written and *what* those authors believed. Choice *E* is incorrect; this is the opposite of the correct answer, the notion that science fiction was not taken seriously.

17. C: V.t. means the verb is transitive (v.t. is short for the Latin words *verbum transitivum*).

18. D: The hyphen in *involve* is present to indicate how many syllables are present. When people pronounce the word *involve*, there's a break between the *in* and *volve*. The hyphen is there to indicate that break.

19. E: Latin is the oldest origins for the word *involve*. The origins can be traced back through the entry. In Middle English it was *enoulen*; old French used *involver*, and the most ancient version (Latin) is *involvere*.

20. D: Car 4 gets 25 mpg. This is calculated by taking the miles (200) and dividing by the number of gallons needed (8). For car 1, divide 100 by 4.23; it gets 23.6 mpg. For car 2, divide 50 by 2.083; it gets 24 mpg. For car 3, divide 70 by 3.181; it gets 22 mpg. Car 3 gets the worst gas mileage.

21. C: See number 20 above.

22. C: To get the number of miles a car can drive on one tank, multiply the mpg by the size of the tank.

For car 1, multiply 23.6 mpg x 12-gallon tank = 283.2 miles
For car 2, 24 mpg x 13-gallon tank = 312 miles
For car 3, 22 mpg x 16-gallon tank = 352 miles
For car 4, 25 mpg x 14-gallon tank = 350 miles.

Therefore, car 3 can travel the greatest distance on one tank of gas. Car 1 can travel the least amount with a range of only 283.2 miles.

23. A: See number 22 above.

24. C: Throughout the entire text, the author maintains a persuasive tone. He argues that punishment should quickly follow the crime and gives a host of reasons why: it's more humane; it helps the prisoner to understand the nature of his or her crimes; it makes a better example for society. To confirm it's a persuasive stance, try reversing the argument. If the position cannot be reversed, then it's not persuasive. In this instance, the reader could argue in rebuttal that the punishment does not have to quickly follow the crime. Regardless of the veracity of this argument, simply creating it proves that the passage is persuasive.

25. C: This passage was written with a cause/effect structure. The cause is that the length between incarceration and trial should be as short as possible. The author, then, lists multiple effects of this cause. There are several key words that indicate this is a cause/effect argument. For instance, the author states, "The degree of the punishment, and the consequences of a crime, ought to be so contrived, as to have the greatest possible effect on others, with the least possible pain to the delinquent." The key words *as to have* indicate that changing the manner of punishment will change the outcome. Similarly, the authors states, "An immediate punishment is more useful; because the smaller the interval of time between the punishment and the crime, the stronger and more lasting will be the association of the two ideas of Crime and Punishment." Similar to *as to have* in the previous excerpt, *because* shows causation. In this instance, the author argues that the shorter the duration between crime and punishment, the more criminals will grasp the consequences of their actions. In general, for cause-effect passages, keep a lookout for words like *because*, *since*, *consequently*, *so*, and *as a result*.

26. D: "The more immediately after the commission of a crime a punishment is inflicted, the more just and useful it will be," best exemplifies the main idea of this passage. All subsequent discussion links back to this main idea and plays the role of supporting details. "The vulgar, that is, all men who have no general ideas or universal principles, act in consequence of the most immediate and familiar associations," supports this idea because the "vulgar," criminals, in other words, are used to making quick associations and are not used to delaying gratification or ignoring their impulses. "Crimes of less importance are commonly punished, either in the obscurity of a prison, or the criminal is transported, to give, by his slavery, an example to societies which he never offended," supports the main idea because the author argues that this is the wrong way to punish because if the punishment occurs in the same area the crime was committed, then the punishment will have more effect, since criminals will associate the areas with their crimes.

Furthermore, transferring a prisoner takes time and delays punishment. For Choice *C*, the author states, "Men do not, in general, commit great crimes deliberately, but rather in a sudden gust of passion; and they commonly look on the punishment due to a great crime as remote and improbable." To reduce the sense of punishments being remote and improbable, criminals must, according to the author, receive an immediate punishment. Therefore, by removing a lengthy gap between crime and punishment, a criminal's punishment will be close and probable, which the author argues is the most humane way to punish and the mark of a civilized society. Choice *E* states "The public punishment, therefore, of small crimes will make a greater impression, and, by deterring men from the smaller, will effectually prevent the greater." This sentence is more of an add-on to the main idea, not the main idea itself.

27. A: The author would disagree most strongly with the statement *criminals are incapable of connecting crime and punishment*. Though the author states that criminals are often passionate and consider punishment unlikely in the heat of the crime, the entire premise of the passage is that reducing the time between crime and punishment increases the likelihood of an association. He also argues that if a society does this consistently, the probability that individuals will consider the consequences of their actions increases. *A punishment should quickly follow the crime* is a restatement of the main idea, supported by evidence throughout the passage. *Most criminals do not think about the consequences of their actions*. Though the author makes this clear, he goes on to say that, in general, reducing the time between crime and punishment will have the most positive effect on the prisoner and on society. *Where a criminal is being punished is just as important as when*. The author argues in the passage that a punishment should be immediate and near where the crime originally occurred. The author also argues that offenders should be punished publicly for small crimes.

28. D: Cellular phone 4 offers the best value for two months. Since there's no cost for the cell phone, take \$65 X 2 = \$130.

Cellular phone 1: \$200 + \$35 x 2 = \$270
Cellular phone 2: \$100 + \$25 x 2 = \$150
Cellular phone 3: \$50 + \$45 x 2 = \$140

29. B: Cellular phone 2 offers the best value.

Cellular phone 2: \$100 + \$25 x 12 = \$400
Cellular phone 1: \$200 + \$35 x 12 = \$620
Cellular phone 3: \$50 + \$45 x 12 = \$590
Cellular phone 4: \$65 x 12 = \$744

30. B: Cellular phone 2 still offers the best value: \$100 + (\$25 + \$30) x 12 = \$760.

Cellular phone 1: \$200 + (\$35 + \$20) x 12 = \$860
Cellular phone 3: \$50 + (\$45 + \$25) x 12 = \$890
Cellular phone 4: (\$65 + \$7) x 12 = \$864.

31. C: Brand-X Management uses a proven technique to deliver bad news: sandwich it between good. The first paragraph extolls the virtue of its employees: "Brand-X wouldn't be where it is today without the hard work of countless, unsung heroes and is truly blessed to have such an amazing staff." The management of Brand-X is proud of its employees. The second paragraph delivers the bad news. Along with the announcement of the termination of the childcare program, Brand-X management provides several reasons why, such as they're "subject to the same economic forces as any other company" and their decision "allowed Brand-X to still stay competitive while not reducing staff." The passage implies that they didn't want to—they had to. The tone of the second paragraph is apologetic. The third paragraph ends on a positive note. Management reminds staff that other services—employee rewards, vacation and sick days, retirement options, volunteer days—have not been impacted, and on Friday employees will have a luncheon and be allowed to wear jeans. This passage is there to not only remind employees why it's still beneficial to work there, but also to provide evidence that Brand-X wants its employees to be happy by hosting luncheons and letting employees wear jeans. It's designed to reduce their hostilities about cancelling the program, or, in other words, placate or appease them.

32. B: "Unfortunately, Brand-X is subject to the same economic forces as any other company." This line was most likely provided to redirect blame from Brand-X to outside forces. Throughout the second paragraph, Brand-X consistently defends its decision to cancel the childcare program. Based on the context of the paragraph, it's continually implied that management was left with no choice. While it might be true that Brand-X is no worse than other companies (others are forced to do the same), this statement supports the idea that it's beyond their control, a message reiterated throughout the paragraph. Nowhere in the paragraph does Brand-X encourage employees to look for work elsewhere. In fact, management does the opposite by reminding employees why they want to continue working there: employee rewards, vacation and sick days, retirement options, and volunteer days. The tone of the passage is not intimidating. No threats are made, implied or otherwise. For example, when management makes reference to reducing staff, it's described as an undesirable choice, not a viable option. The Brand-X memo does not attempt to insult itself, so Choice *E* is incorrect.

33. C: In modern formatting, longer works, such as books, are indicated through the use of italics. Shorter works, such as poems or short stories, are indicated through the use of quotation marks.

Accordingly, *The Sound and the Fury*, *For Whom the Bell Tolls*, and *The Grapes of Wrath* are full-length books, while "A&P" and "Barn Burning" are short stories.

34. B: See 33 above.

35. E: Based on what Carl has encountered—a spider web and a mouse—he would be hesitant, lest he encounter another critter. *Fearfully* is too strong a word (unless Carl has a phobia) and would indicate something more serious, like mortal danger. Nowhere in the passage does it indicate Carl is angry, and for Carl to be suspicious, he would have to assume that someone in his life was trying to deceive him. There's no evidence of deception here. *Excitingly* does not fit the context here because Carl has run into one unfortunate event after another.

36. C: *Initiative* would be the best replacement for *gumption*. Both words indicate a strong desire to complete a task and overcome obstacles. *Stubbornness* would imply that Jason was resistant to changing his behavior despite it being good or beneficial for him. Despite Jason being lazy, at the end of the passage, he finds the intrinsic motivation necessary to pass the class. The word *ambition* indicates not only a desire for success but also a desire to gain power and influence. Though Jason eventually wants to succeed, he's not trying to gain power or influence. *Courage* is close, but it's not quite as relevant as the word *initiative* here.

37. C: To arrive at the correct answer, first calculate the actual times Lucy and Bob could arrive at and depart from the dock. Lucy needs an additional thirty minutes, so that equates to 1:15 p.m., not 12:45. Bob works at 6:30 p.m. and needs an hour subtracted to get to work on time, so that means he would have to leave the dock by 5:30 p.m. Ship 3 leaves at 1:30 p.m., which gives Lucy an additional fifteen minutes' leeway, then returns at 5:30 p.m., which gives Bob the hour he needs to arrive at work on time. Ships 1 and 2 depart at 1:10 p.m. and 1:00 p.m., respectively. Both departure times are too early for Lucy. Ship 4 returns at 5:45 p.m., which is later than Bob can leave the dock for work.

38. D: To arrive at the correct answer, subtract the return times from the arrival times. Assuming it takes approximately the same amount of time for each ship to return, the biggest number should indicate the most amount of time spent on each respective island. Ship 4 provides the most amount of time with 3 hours, 55 minutes. Ship 1 allows 2 hours, 10 minutes; ship 2 is 2 hours, 45 minutes, and ship 3 is 2 hours, 50 minutes.

39. C: Despite the opposite stances in Passages 1 and 2, both authors establish that cell phones are a strong part of culture. In Passage 1 the author states, "Today's student is already strongly rooted in technology." In Passage 2 the author states, "Students are comfortable with cell phones." The author of Passage 2 states that cell phones have a "time and place." The author of Passage 2 would disagree with the statement that *teachers should incorporate cell phones into curriculum whenever possible. Cell phones are useful only when an experienced teacher uses them properly*—this statement is implied in Passage 2, but the author in Passage 1 says cell phones are "indispensable." In other words, no teacher can do without them. *Despite a good lesson plan, cell phone disruptions are impossible to avoid.* This is not supported by either passage. Even though the author in the second passage is more cautionary, the author states, "This can prove advantageous if done correctly." Therefore, there is a possibility that a classroom can run properly with cell phones. *Cell phones are necessary in an emergency.* The author of Passage 1 would agree with this statement, but the author of Passage 2 does not think cell phones are appropriate in a classroom setting.

40. D: Here, "speaking" has nothing to do with communication. The original quotation from Passage 1 reads, *Since today's student is already strongly rooted in technology, when teachers incorporate cell* *phones, they're 'speaking' the student's language*. The first dependent clause makes reference to technology, then uses the word "speaking" as an analogy. The "language" the author refers to is culture. The author never intended the word "speak" to imply actual communication. *Cell phones are a free, readily available resource*; *cell phones incorporate other forms of technology*; and *due to their mobility, cell phones are excellent in an emergency* are all listed as reasons in Passage 1 to have cell phones. *Cell phones allow higher engagement between students and teachers.* This sentiment is represented toward the end of Passage 1.

41. B: The author states in Passage 2, "Confiscating phones is akin to rummaging through students' backpacks." The author most likely included this passage to exemplify how strongly students believe this is a right. This quotation is the evidence or example, and it ties to the previous sentence: "This type of disrespectful behavior is often justified by the argument that cell phones are not only a privilege but also a right." The backpack example illustrates how students believe it's a right. *Indicate how unlikely students are to change their minds*; *exemplify how easily the modern student is offended*; *demonstrate how illogical most students' beliefs are*; and *explain how the rules of education are rapidly changing*; are not supported by Passage 2.

42. E: The best substitute for *surreptitiously* would be *slyly*. Let's look at the context of the sentence: "a sizable percentage of a classroom pretends to pay attention while surreptitiously playing on their phones." The key word here is *pretends*. The students in the classroom, therefore, want to use their cell phones but are afraid of getting caught. *Privately* and *casually* don't work because they leave out the connotation of hiding a wrong act or ulterior motive. *Obstinately* and *defiantly* are also incorrect since students are hiding their disobedience.

43. C: Safflower oil has a smoking point of 510° Fahrenheit. Since Zack's recipe doesn't exceed 430° F, it won't reach the smoking point. In addition, safflower oil doesn't contain peanuts, and safflower oil has a neutral taste. Though clarified butter has a high smoke point, too, it doesn't have a neutral taste. Zack is allergic to peanut oil. Lard, at 374° F, has too low a smoking point and doesn't have a neutral taste. Coconut oil has a smoking point of 350° F and doesn't have a neutral taste.

44. A: Clarified butter has a high smoke point (485°), which would be well above 430° F and doesn't have a neutral taste, which in this scenario Zack prefers. Here, peanut oil is a neutral flavor. Lard, at 374° F, has too low a smoking point. Safflower oil has a smoking point of 510° F, but, here, the flavor is neutral, and he wants a flavored oil. Coconut oil is not neutral, but the smoking point of 350° F is too low.

45. B: *Abusive* would be the best substitute for *vituperative.* Based on the context clues of "long day of work," "fists clenched," and "eyebrows knitted," Julio's mother is angry with her son. Therefore, one can infer that her language will match her nonverbal behavior. *Annoyed* is simply too mild of a word and does not match with her enraged posture. *Passionate* is too vague of a word that can be matched to a different number of moods. *Sorrowful* contradicts the angry posture she's adopted. *Virtuous* means innocent so this is incorrect.

46. E: This passage is designed to *persuade*. The whole purpose of the passage is to convince vacationers to come to Random Lake. There are some informative aspects of the passage, such as what's available – boating, house rentals, pontoon boats, but an extremely positive spin is put on the area, a spin that's designed to attract visitors. To prove it's persuasive, the argument can be reversed: Random Lake is *not* a good place to vacation. With a lack of adjectives and adverbs, there's a lack of *descriptive* detail, and this passage doesn't delight or *entertain* like a witty narrative might.

47. C: "Pontoon boats, kayaks, and paddle boats are available for rental" is out of place. The sentence before refers to fully furnished homes, and the sentence after refers to new developments. This sentence would have fit much better near, "Swimming, kayaking, boating, fishing, volleyball, mini-golf, go-cart track, eagle watching, nature trails," because the activities of the area are described. "This summer, why not rent a house on Random Lake?" is followed, logically, by a description of where Random Lake is. "There are enough activities there to keep a family busy for a month," is preceded by a listing of all those activities. "With several new housing developments slated for grand opening in March, the choices are endless!" links to "furnished houses," because this is an opportunity to find even more houses. "With the Legends and the Broadmoor developments slated for grand openings in March, the choices are endless!" This is an effective conclusion to the passage, so Choice *E* is not out of place.

48. C: *The Random Lake area is growing.* This can be deduced with the passage, "With the Legends and The Broadmoor developments slated for grand openings in March, the choices are endless!" Considering buildings are being added, not razed, the Random Lake area would have to be growing. *Random City is more populated than Random Lake.* The word *city* is not necessarily indicative of size. There's simply not enough information to determine the size of either Random City or Random Lake. *The Random Lake area is newer than Random City.* Though this might seem logical with the addition of new buildings at Random Lake, there's no way to confirm this. The reader simply doesn't know enough about Random City to draw a comparison. *Random Lake prefers families to couples.* The ad definitely appeals to families but, "There are hotels available for every lifestyle and budget," proves that Random Lake is trying to appeal to everyone. Furthermore, "Prefer quiet? Historical? Modern?" proves that there are accommodations for every lifestyle. *You can go kayaking at a Random Lake Hotel.* This is apparent in the second passage where it names different activities for people who stay at Random Lake hotels.

49. A: The tone is exasperated. While contemplative is an option because of the inquisitive nature of the text, Choice *A* is correct because the speaker is annoyed by the thought of being included when he felt that the fellow members of his race were being excluded. The speaker is not nonchalant, nor accepting of the circumstances which he describes. The speaker also does not come across as irate, as that would alienate the audience.

50. E: Choice *C*, *contented*, is the only word that has a different meaning. Furthermore, the speaker expresses objection and disdain throughout the entire text.

Mathematics

1. B: 15,412

Set up the problem and add each column, starting on the far right (ones). Add, carrying anything over 9 into the next column to the left. Solve from right to left.

2. A: Using the order of operations, multiplication and division are computed first from left to right. Multiplication is on the left; therefore, the teacher should perform multiplication first.

3. E: A factor of 36 is any number that can be divided into 36 and have no remainder. $36 = 36 \times 1, 18 \times 2, 9 \times 4, \text{and } 6 \times 6$. Therefore, it has 7 unique factors: 36, 18, 9, 6, 4, 2, and 1.

4. B: A number is prime because its only factors are itself and 1. Positive numbers (greater than zero) can be prime numbers.

5. C: $\frac{31}{36}$

Set up the problem and find a common denominator for both fractions.

$$\frac{5}{12} + \frac{4}{9}$$

Multiply each fraction across by 1 to convert to a common denominator.

$$\frac{5}{12} \times \frac{3}{3} + \frac{4}{9} \times \frac{4}{4}$$

Once over the same denominator, add across the top. The total is over the common denominator.

$$\frac{15 + 16}{36} = \frac{31}{36}$$

6. A: Order of operations follows PEMDAS—Parentheses, Exponents, Multiplication and Division from left to right, and Addition and Subtraction from left to right.

7. C: 0.63

Divide 5 by 8, which results in 0.625. This rounds up to 0.63.

8. E: 8,685

Set up the problem, with the larger number on top. Begin subtracting with the far-right column (ones). Borrow 10 from the column to the left, when necessary.

9. A: The place value to the right of the thousandth place, which would be the ten-thousandth place, is what gets used. The value in the thousandth place is 7. The number in the place value to its right is greater than 4, so the 7 gets bumped up to 8. Everything to its right turns to a zero, to get 245.2680. The zero is dropped because it is part of the decimal.

10. E: 37.797

Set up the problem, larger number on top and numbers lined up at the decimal. Begin subtracting with the far-right column. Borrow 10 from the column to the left, when necessary.

11. C: Numbers should be lined up by decimal places before subtraction is performed. This is because subtraction is performed within each place value. The other operations, such as multiplication, division, and exponents (which is a form of multiplication), involve ignoring the decimal places at first and then including them at the end.

12. C: $\frac{19}{24}$

Set up the problem and find a common denominator for both fractions.

$$\frac{23}{24} - \frac{1}{6}$$

Multiply each fraction across by 1 to convert to a common denominator.

$$\frac{23}{24} \times \frac{1}{1} - \frac{1}{6} \times \frac{4}{4}$$

Once over the same denominator, subtract across the top.

$$\frac{23 - 4}{24} = \frac{19}{24}$$

13. D: $\frac{2}{9}$

Set up the problem and find a common denominator for both fractions.

$$\frac{43}{45} - \frac{11}{15}$$

Multiply each fraction across by 1 to convert to a common denominator.

$$\frac{43}{45} \times \frac{1}{1} - \frac{11}{15} \times \frac{3}{3}$$

Once over the same denominator, subtract across the top.

$$\frac{43 - 33}{45} = \frac{10}{45}$$

Reduce.

$$\frac{10 \div 5}{45 \div 5} = \frac{2}{9}$$

14. B: $\frac{14}{25}$

Since 0.56 goes to the hundredths place, it can be placed over 100:

$$\frac{56}{100}$$

Essentially, the way we got there is by multiplying the numerator and denominator by 100:

$$\frac{0.56}{1} \times \frac{100}{100} = \frac{56}{100}$$

Then, the fraction can be simplified down to 14/25:

$$\frac{56}{100} \div \frac{4}{4} = \frac{14}{25}$$

15. A: 2,504,774

Line up the numbers (the number with the most digits on top) to multiply. Begin with the right column on top and the right column on bottom.

Move one column left on top and multiply by the far-right column on the bottom. Remember to add the carry over after you multiply. Continue that pattern for each of the numbers on the top row.

Starting on the far-right column on top repeat this pattern for the next number left on the bottom. Write the answers below the first line of answers; remember to begin with a zero placeholder. Continue for each number in the top row.

Starting on the far-right column on top, repeat this pattern for the next number left on the bottom. Write the answers below the first line of answers. Remember to begin with zero placeholders.

Once completed, ensure the answer rows are lined up correctly, then add.

16. A: The first step is to divide up $150 into four equal parts. 150/4 is 37.5, so she needs to save an average of $37.50 per day.

17. E: The teacher would be introducing fractions. If a pie was cut into 6 pieces, each piece would represent $\frac{1}{6}$ of the pie. If one piece was taken away, $\frac{5}{6}$ of the pie would be left over.

18. A: $\frac{810}{2921}$

Line up the fractions.

$$\frac{15}{23} \times \frac{54}{127}$$

Multiply across the top and across the bottom.

$$\frac{15 \times 54}{23 \times 127} = \frac{810}{2921}$$

19. D: There are no millions, so the millions period consists of all zeros. 182 is in the billions period, 36 is in the thousands period, 421 is in the hundreds period, and 356 is the decimal.

20. D: $11\frac{3}{4}$

Set up the division problem.

$$16\overline{)188}$$

16 does not go into 1 but does go into 18 so start there.

$$\begin{array}{r} 11 \\ 16\overline{)188} \\ -16 \\ \hline 28 \\ -16 \\ \hline 12 \end{array}$$

The result is $11\frac{12}{16}$

Reduce the fraction for the final answer.

$$11\frac{3}{4}$$

21. D: Division can be computed as a repetition of subtraction problems by subtracting multiples of 24.

22. D:

$$\frac{300}{80} = \frac{30}{8}$$

$$\frac{300}{80} = \frac{30}{8} = \frac{15}{4} = 3.75$$

But Bernard is only working full days, so he will need to work 4 days, since 3 days are not sufficient.

23. E: The value went up by $165,000 – $150,000 = $15,000. Out of $150,000, this is:

$$\frac{15,000}{150,000} = \frac{1}{10}$$

Convert this to having a denominator of 100, the result is $\frac{10}{100}$ or 10%.

24. B: A number raised to an exponent is a compressed form of multiplication. For example:

$$10^3 = 10 \times 10 \times 10$$

25. B: The perimeter of a rectangle is the sum of all four sides. Therefore, the answer is:

$$P = 14 + 8\frac{1}{2} + 14 + 8\frac{1}{2}$$

$$14 + 14 + 8 + \frac{1}{2} + 8 + \frac{1}{2} = 45 \text{ square inches}$$

26. C: Kilograms, grams, kilometers, and meters. Inches, pounds, and baking measurements, such as tablespoons, are not part of the metric system. Kilograms, grams, kilometers, and meters are part of the metric system.

27. E: It shows the associative property of multiplication. The order of multiplication does not matter, and the grouping symbols do not change the final result once the expression is evaluated.

28. B: $12 \times 750 = 9{,}000$. Therefore, there are 9,000 milliliters of water, which must be converted to liters. 1,000 milliliters equals 1 liter; therefore, 9 liters of water are purchased.

29. A: According to order of operations, the operation within the parentheses must be completed first. Next, division is completed and then subtraction. Therefore, the expression is evaluated as:

$$(3 + 7) - 6 \div 2 = 10 - 6 \div 2 = 10 - 3 = 7$$

In order to incorrectly obtain 2 as the answer, the operations would have been performed from left to right, instead of following PEMDAS.

30. A: Operations within the parentheses must be completed first. Then, division is completed. Finally, addition is the last operation to complete. When adding decimals, digits within each place value are added together. Therefore, the expression is evaluated as:

$$(2 \times 20) \div (7 + 1) + (6 \times 0.01) + (4 \times 0.001)$$

$$40 \div 8 + 0.06 + 0.004 = 5 + 0.06 + 0.004 = 5.064$$

31. E: The formula for the volume of a rectangular solid would need to be used. The volume of the first box is:

$$V = 3 \times 5 \times 10 = 150 \text{ cubic inches}$$

The second box needs to hold cereal that would take up the same space. The volume of the second box is:

$$V = 5 \times 6 \times h = 30 \times h$$

In order for this to equal 150, *h* must equal 5 inches.

32. C: The measure of two complementary angles sums up to 90 degrees. $90 - 54 = 36$. Therefore, the complementary angle is 36 degrees.

33. B: Instead of multiplying these out, the product can be estimated by using $18 \times 10 = 180$. The error here should be lower than 15, since it is rounded to the nearest integer, and the numbers add to something less than 30.

34. D: The total faculty is 15 + 20 = 35. Therefore, the faculty to student ratio is 35:200. Then, to simplify this ratio, both the numerator and the denominator are divided by 5, since 5 is a common factor of both, which yields 7:40.

35. B: Katie eats $\frac{1}{3}$ of $\frac{3}{4}$ of the pizza. That means she eats $\frac{1}{3} \times \frac{3}{4} = \frac{3}{12} = \frac{1}{4}$ of the pizza. Therefore, $\frac{3}{4} - \frac{1}{4} = \frac{2}{4} = \frac{1}{2}$ of the pizza remains. This problem involves subtraction of fractions with like denominators.

36. A: Taylor's total income is \$20,000 + \$10,000 = \$30,000. 15% of this is $\frac{15}{100} = \frac{3}{20}$.

$$\frac{3}{20} \times \$30{,}000 = \frac{90{,}000}{20} = \frac{9000}{2} = \$4500$$

37. B: Each nickel is worth \$0.05. Therefore, Joshua deposited:

$$12{,}345 \times \$0.05 = \$617.25$$

Working with change is a great way to teach decimals to children, so this problem would be a good introduction to multiplying decimals.

38. E: Since the answer will be in cubic feet rather than inches, the first step is to convert from inches to feet for the dimensions of the box. There are 12 inches per foot, so the box is 24/12 = 2 feet wide, 18/12 = 1.5 feet deep, and 12/12 = 1 foot high. The volume is the product of these three together:

$$2 \times 1.5 \times 1 = 3 \; cubic \; feet$$

39. E: According to the order of operations, multiplication and division must be completed first from left to right. Then, addition and subtraction are completed from left to right. Therefore:

$$9 \times 9 \div 9 + 9 - 9 \div 9$$

$$81 \div 9 + 9 - 9 \div 9$$

$$9 + 9 - 9 \div 9$$

$$9 + 9 - 1$$

$$18 - 1$$

$$17$$

40. B: When giving an answer to a math problem that is in fraction form, it always should be simplified. Both 3 and 15 have a common factor of 3 that can be divided out, so the correct answer is $\frac{3 \div 3}{15 \div 3} = \frac{1}{5}$.

41. A: 13 nurses

Using the given information of 1 nurse to 25 patients and 325 patients, set up an equation to solve for number of nurses (N):

$$\frac{N}{325} = \frac{1}{25}$$

Multiply both sides by 325 to get N by itself on one side.

$$\frac{N}{1} = \frac{325}{25} = 13 \; nurses$$

42. D: 290 beds

Using the given information of 2 beds to 1 room and 145 rooms, set up an equation to solve for number of beds (B):

$$\frac{B}{145} = \frac{2}{1}$$

Multiply both sides by 145 to get B by itself on one side.

$$\frac{B}{1} = \frac{290}{1} = 290 \; beds$$

43. C: x = 150

Set up the initial equation.

$$\frac{2x}{5} - 1 = 59$$

Add 1 to both sides.

$$\frac{2x}{5} - 1 + 1 = 59 + 1$$

Multiply both sides by 5/2.

$$\frac{2x}{5} \times \frac{5}{2} = 60 \times \frac{5}{2} = 150$$

$$x = 150$$

44. C: $51.93

List the givens.

$$Tax = 6.0\% = 0.06$$

$$Sale = 50\% = 0.5$$

$$Hat = \$32.99$$

$$Jersey = \$64.99$$

Calculate the sales prices.

$$Hat\ Sale = 0.5\,(32.99) = 16.495$$

$$Jersey\ Sale = 0.5\,(64.99) = 32.495$$

Total the sales prices.

$$Hat\ sale + jersey\ sale = 16.495 + 32.495 = 48.99$$

Calculate the tax and add it to the total sales prices.

$$Total\ after\ tax = 48.99 + (48.99\ x\ 0.06) = \$51.93$$

45. D: $0.45

List the givens.

$$Store\ coffee\ =\ \$1.23/lbs$$

$$Local\ roaster\ coffee\ =\ \$1.98/1.5\ lbs$$

Calculate the cost for 5 lbs of store brand.

$$\frac{\$1.23}{1\ lbs} \times 5\ lbs\ = \$6.15$$

Calculate the cost for 5 lbs of the local roaster.

$$\frac{\$1.98}{1.5\ lbs} \times 5\ lbs\ = \$6.60$$

Subtract to find the difference in price for 5 lbs.

$$\begin{array}{r} \$6.60 \\ -\$6.15 \\ \hline \$0.45 \end{array}$$

46. D: $3,325

List the givens.

$$1{,}800\ ft. =\ \$2{,}000$$

$$Cost\ after\ 1{,}800\ ft. =\ \$1.00/ft.$$

Find how many feet left after the first 1,800 ft.

$$\begin{array}{lr} & 3{,}125\text{ ft.} \\ - & 1{,}800\text{ ft.} \\ \hline & 1{,}325\text{ ft.} \end{array}$$

Calculate the cost for the feet over 1,800 ft.

$$1{,}325\ ft. \times \frac{\$1.00}{1\ ft}\ = \$1{,}325$$

Total for entire cost.

$$\$2{,}000\ +\ \$1{,}325\ =\ \$3{,}325$$

47. A: 12

Calculate how many gallons the bucket holds.

$$11.4\,L \times \frac{1\,gal}{3.8\,L} = 3\,gal$$

Now how many buckets to fill the pool which needs 35 gallons.

$$35/3 = 11.67$$

Since the amount is more than 11 but less than 12, we must fill the bucket 12 times.

48. D: This problem can be solved by using unit conversion. The initial units are miles per minute. The final units need to be feet per second. Converting miles to feet uses the equivalence statement 1 mile = 5,280 feet. Converting minutes to seconds uses the equivalence statement 1 minute = 60 seconds. Setting up the ratios to convert the units is shown in the following equation:

$$\frac{72\,miles}{90\,minutes} \times \frac{1\,minute}{60\,seconds} \times \frac{5280\,feet}{1\,mile} = 70.4 \text{ feet per second}$$

The initial units cancel out, and the new units are left.

49. B: 35.1%

To convert from a decimal to a percentage, the decimal needs to be moved two places to right. In this case, that makes 0.351 become 35.1%.

50. A: 22%

Converting from a fraction to a percentage generally involves two steps. First, the fraction needs to be converted to a decimal.

Divide 2 by 9 which results in $0.\overline{22}$. The top line indicates that the decimal actually goes on forever with an endless amount of 2's.

Second, the decimal needs to be moved two places to the right:

$$22\%$$

CBEST Practice Test #3

Reading

Questions 1–4 are based on the following passage.

Smoking tobacco products is terribly destructive. A single cigarette contains over 4,000 chemicals, including 43 known carcinogens and 400 deadly toxins. Some of the most dangerous ingredients include tar, carbon monoxide, formaldehyde, ammonia, arsenic, and DDT. Smoking can cause numerous types of cancer including throat, mouth, nasal cavity, esophageal, gastric, pancreatic, renal, bladder, and cervical cancer.

Cigarettes contain a drug called nicotine, one of the most addictive substances known to man. Addiction is defined as a compulsion to seek the substance despite negative consequences. According to the National Institute of Drug Abuse, nearly 35 million smokers expressed a desire to quit smoking in 2015; however, more than 85 percent of those who struggle with addiction will not achieve their goal. Almost all smokers regret picking up that first cigarette. You would be wise to learn from their mistake if you have not yet started smoking.

According to the U.S. Department of Health and Human Services, 16 million people in the United States presently suffer from a smoking-related condition and nearly nine million suffer from a serious smoking-related illness. According to the Centers for Disease Control and Prevention (CDC), tobacco products cause nearly six million deaths per year. This number is projected to rise to over eight million deaths by 2030. Smokers, on average, die ten years earlier than their nonsmoking peers.

In the United States, local, state, and federal governments typically tax tobacco products, which leads to high prices. Nicotine users who struggle with addiction sometimes pay more for a pack of cigarettes than for a few gallons of gas. Additionally, smokers tend to stink. The smell of smoke is all-consuming and creates a pervasive nastiness. Smokers also risk staining their teeth and fingers with yellow residue from the tar.

Smoking is deadly, expensive, and socially unappealing. Clearly, smoking is not worth the risks.

1. Which of the following best describes the passage?
   a. Narrative
   b. Persuasive
   c. Expository
   d. Technical
   e. Informative

2. Which of the following statements most accurately summarizes the passage?
    a. Almost all smokers regret picking up that first cigarette.
    b. Tobacco is deadly, expensive, and socially unappealing, and smokers would be much better off kicking the addiction.
    c. In the United States, local, state, and federal governments typically tax tobacco products, which leads to high prices.
    d. Tobacco products shorten smokers' lives by ten years and kill more than six million people per year.
    e. Tobacco is less healthy than many alternatives.

3. The author would be most likely to agree with which of the following statements?
    a. Smokers should only quit cold turkey and avoid all nicotine cessation devices.
    b. Other substances are more addictive than tobacco.
    c. Smokers should quit for whatever reason that gets them to stop smoking.
    d. People who want to continue smoking should advocate for a reduction in tobacco product taxes.
    e. Smokers don't have the desire to quit and often don't see their smoking as a bad habit.

4. Which of the following represents an opinion statement on the part of the author?
    a. According to the Centers for Disease Control and Prevention (CDC), tobacco products cause nearly six million deaths per year.
    b. Nicotine users who struggle with addiction sometimes pay more for a pack of cigarettes than a few gallons of gas.
    c. They also risk staining their teeth and fingers with yellow residue from the tar.
    d. Additionally, smokers tend to stink. The smell of smoke is all-consuming and creates a pervasive nastiness.
    e. Smokers, on average, die ten years earlier than their nonsmoking peers.

Questions 5–7 are based on the following passage.

> George Washington emerged out of the American Revolution as an unlikely champion of liberty. On June 14, 1775, the Second Continental Congress created the Continental Army, and John Adams, serving in the Congress, nominated Washington to be its first commander. Washington fought under the British during the French and Indian War, and his experience and prestige proved instrumental to the American war effort. Washington provided invaluable leadership, training, and strategy during the Revolutionary War. He emerged from the war as the embodiment of liberty and freedom from tyranny.
>
> After vanquishing the heavily favored British forces, Washington could have pronounced himself as the autocratic leader of the former colonies without any opposition, but he famously refused and returned to his Mount Vernon plantation. His restraint proved his commitment to the fledgling state's republicanism. Washington was later unanimously elected as the first American president. But it is Washington's farewell address that cemented his legacy as a visionary worthy of study.
>
> In 1796, President Washington issued his farewell address by public letter. Washington enlisted his good friend, Alexander Hamilton, in drafting his most famous address. The letter expressed Washington's faith in the Constitution and rule of law. He encouraged his fellow Americans to put aside partisan differences and establish a national union. Washington warned Americans

against meddling in foreign affairs and entering military alliances. Additionally, he stated his opposition to national political parties, which he considered partisan and counterproductive.

Americans would be wise to remember Washington's farewell, especially during presidential elections when politics hits a fever pitch. They might want to question the political institutions that were not planned by the Founding Fathers, such as the nomination process and political parties themselves.

5. Which of the following statements is logically based on the information contained in the passage above?

a. George Washington's background as a wealthy landholder directly led to his faith in equality, liberty, and democracy.
b. George Washington would have opposed America's involvement in the Second World War.
c. George Washington would not have been able to write as great a farewell address without the assistance of Alexander Hamilton.
d. George Washington would probably not approve of modern political parties.
e. George Washington would likely befriend former President Barack Obama.

6. Which of the following statements is the best description of the author's purpose in writing this passage about George Washington?

a. To caution American voters about being too political during election times because George Washington would not have agreed with holding elections.
b. To introduce George Washington to readers as a historical figure worthy of study
c. To note that George Washington was more than a famous military hero
d. To convince readers that George Washington is a hero of republicanism and liberty
e. To inform American voters about a Founding Father's sage advice on a contemporary issue and explain its applicability to modern times

7. In which of the following materials would the author be the most likely to include this passage?

a. A history textbook
b. An obituary
c. A fictional story
d. A newspaper editorial
e. A research journal

Questions 8–12 are based on the following passage.

Christopher Columbus is often credited for discovering America. This is incorrect. First, it is impossible to "discover" something where people already live; however, Christopher Columbus did explore places in the New World that were previously untouched by Europe, so the term "explorer" would be more accurate. Another correction must be made, as well: Christopher Columbus was not the first European explorer to reach the present day Americas! Rather, it was Leif Erikson who first came to the New World and contacted the natives, nearly five hundred years before Christopher Columbus.

Leif Erikson, the son of Erik the Red (a famous Viking outlaw and explorer in his own right), was born in either 970 or 980, depending on which historian you seek. His own family, though, did not raise Leif, which was a Viking tradition. Instead, one of Erik's prisoners taught Leif reading and writing, languages, sailing, and weaponry. At age 12, Leif was considered a man and returned to his family. He killed a man during a dispute shortly after his return, and the council banished the Erikson clan to Greenland.

In 999, Leif left Greenland and traveled to Norway where he would serve as a guard to King Olaf Tryggvason. It was there that he became a convert to Christianity. Leif later tried to return home with the intention of taking supplies and spreading Christianity to Greenland, however his ship was blown off course and he arrived in a strange new land: present day Newfoundland, Canada".

When he finally returned to his adopted homeland Greenland, Leif consulted with a merchant who had also seen the shores of this previously unknown land we now know as Canada. The son of the legendary Viking explorer then gathered a crew of 35 men and set sail. Leif became the first European to touch foot in the New World as he explored present-day Baffin Island and Labrador, Canada. His crew called the land Vinland since it was plentiful with grapes.

During their time in present-day Newfoundland, Leif's expedition made contact with the natives whom they referred to as Skraelings (which translates to "wretched ones" in Norse). There are several secondhand accounts of their meetings. Some contemporaries described trade between the peoples. Other accounts describe clashes where the Skraelings defeated the Viking explorers with long spears, while still others claim the Vikings dominated the natives. Regardless of the circumstances, it seems that the Vikings made contact of some kind. This happened around 1000, nearly five hundred years before Columbus famously sailed the ocean blue.

Eventually, in 1003, Leif set sail for home and arrived at Greenland with a ship full of timber.

In 1020, seventeen years later, the legendary Viking died. Many believe that Leif Erikson should receive more credit for his contributions in exploring the New World.

8. Which of the following best describes how the author generally presents the information?
   a. Chronological order
   b. Comparison-contrast
   c. Cause-effect
   d. Conclusion-premises
   e. Spatial order

9. Which of the following is an opinion, rather than historical fact, expressed by the author?
   a. Leif Erikson was definitely the son of Erik the Red; however, historians debate the year of his birth.
   b. Leif Erikson's crew called the land Vinland since it was plentiful with grapes.
   c. Leif Erikson deserves more credit for his contributions in exploring the New World.
   d. Leif Erikson explored the Americas nearly five hundred years before Christopher Columbus.
   e. Leif's expedition made contact with the natives whom they referred to as Skraelings.

10. Which of the following most accurately describes the author's main conclusion?
   a. Leif Erikson is a legendary Viking explorer.
   b. Leif Erikson deserves more credit for exploring America hundreds of years before Columbus.
   c. Spreading Christianity motivated Leif Erikson's expeditions more than any other factor.
   d. Leif Erikson contacted the natives nearly five hundred years before Columbus.
   e. Leif Erikson discovered the Americas.

11. Which of the following best describes the author's intent in the passage?
    a. To entertain
    b. To inform
    c. To alert
    d. To suggest
    e. To share

12. Which of the following can be logically inferred from the passage?
    a. The Vikings disliked exploring the New World.
    b. Leif Erikson's banishment from Iceland led to his exploration of present-day Canada.
    c. Leif Erikson never shared his stories of exploration with the King of Norway.
    d. Historians have difficulty definitively pinpointing events in the Vikings' history.
    e. Christopher Columbus knew of Leif Erikson's explorations

Questions 13–17 are based on the chart following a brief introduction to the topic.

The American Civil War was fought from 1861 to 1865. It is the only civil war in American history. While the South's secession was the initiating event of the war, the conflict grew out of several issues like slavery and differing interpretations of individual state rights. General Robert E. Lee led the Confederate Army for the South for the duration of the conflict (although other generals held command positions over individual battles, as you will see next). The North employed a variety of lead generals, but Ulysses S. Grant finished the war as the victorious general. There were more American casualties in the Civil War than any other military conflict in American history.

Civil War Casualties by Battle (approximate)					
Battle	**Date**	**Union General**	**Confederate General**	**Union Casualties**	**Confederate Casualties**
Gettysburg	July 1863	George Meade	Robert E. Lee	23,049	28,063
Chancellorsville	May 1863	Joseph Hooker	Robert E. Lee	17,304	13,460
Shiloh	April 1862	Ulysses S. Grant	Albert Sydney Johnston	13,047	10,669
Cold Harbor	May 1864	Ulysses S. Grant	Robert E. Lee	12,737	4,595
Atlanta	July 1864	William T. Sherman	John Bell Hood	3,722	5,500

13. In which of the following battles were there more Confederate casualties than Union casualties?
    a. The one in May 1864
    b. Chancellorsville
    c. The one in April 1862
    d. Shiloh
    e. Atlanta

14. Which one of the following battles occurred first?
    a. Cold Harbor
    b. Chancellorsville
    c. Atlanta
    d. Shiloh
    e. Gettysburg

15. Robert E. Lee did not lead the Confederate forces in which one of the following battles?
   a. Atlanta
   b. The one in May 1863
   c. Cold Harbor
   d. Gettysburg
   e. The one in May 1864

16. In which of the following battles did the Union casualties exceed the Confederate casualties by the greatest number?
   a. Cold Harbor
   b. Chancellorsville
   c. Atlanta
   d. Shiloh
   e. Gettysburg

17. The total number of American casualties suffered at the battle of Gettysburg is about double the total number of casualties suffered at which one of the following battles?
   a. Cold Harbor
   b. Chancellorsville
   c. Atlanta
   d. Shiloh
   e. Cannot be determined with the information provided

Questions 18 and 19 are based on the graphic that follows a brief introduction to the topic.

The United States Constitution directs Congress to conduct a census of the population to determine the country's population and demographic information. The United States Census Bureau carries out the survey. In 1790, then Secretary of State Thomas Jefferson conducted the first census, and the most recent U.S. census was in 2010. The next U.S. census will be the first to be issued primarily through the Internet.

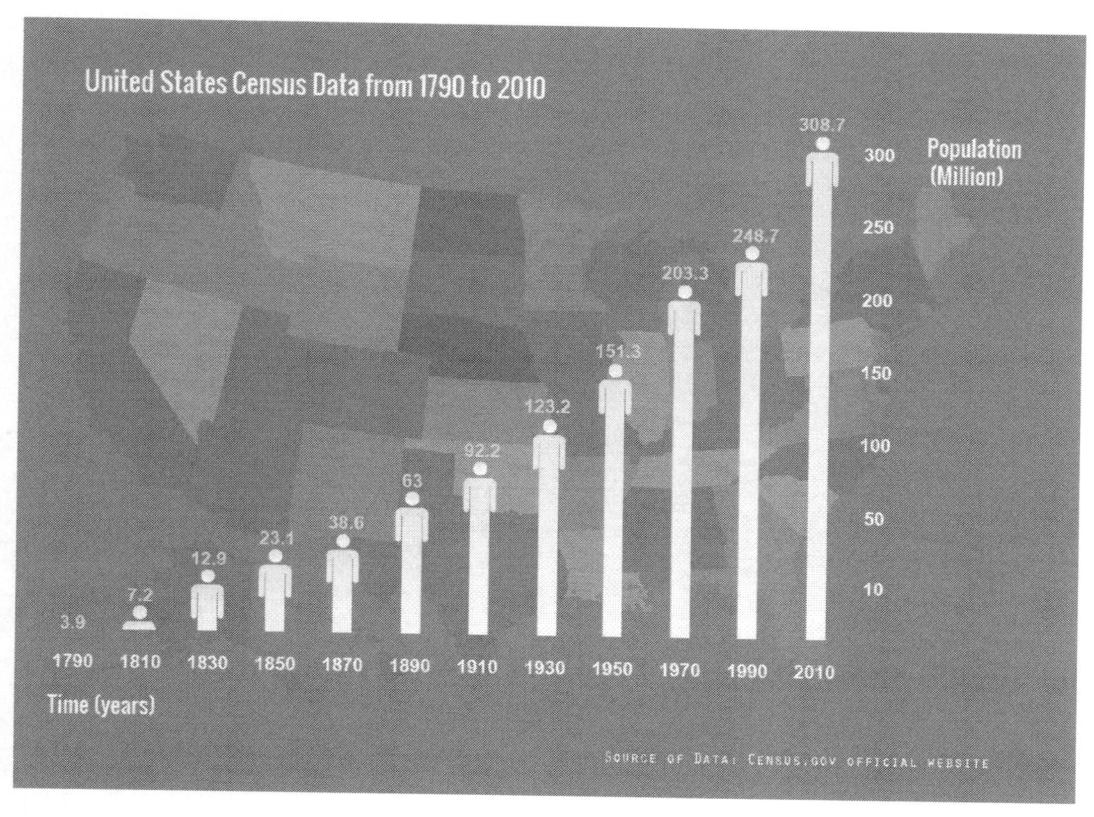

18. In which of the following years was the United States population less than it was in 1930?
    a. 1950
    b. 1970
    c. 2010
    d. 1990
    e. 1910

19. In what year did the population increase the most during a twenty-year interval?
    a. From 1930 to 1950
    b. From 1950 to 1970
    c. From 1970 to 1990
    d. From 1990 to 2010
    e. From 1790 to 1810

Questions 20–22 are based on the graphic following a brief introduction to the topic.

A food chain is a diagram used by biologists to better understand ecosystems. It represents the interrelationships between different plants and animals. The energy is derived from the sun and converted into stored energy by plants through photosynthesis, which travels up the food chain. The energy returns to the ecosystem after the organisms die and decompose back into the Earth. This process is an endless cycle.

In food chains, living organisms are grouped into categories called primary producers and consumers, which come in multiple tiers. For example, secondary consumers feed on primary consumers, while tertiary consumers feed on secondary consumers. Apex predators are the animals at the top of the food chain. They are the highest category consumer in an ecosystem, and apex predators do not have natural predators.

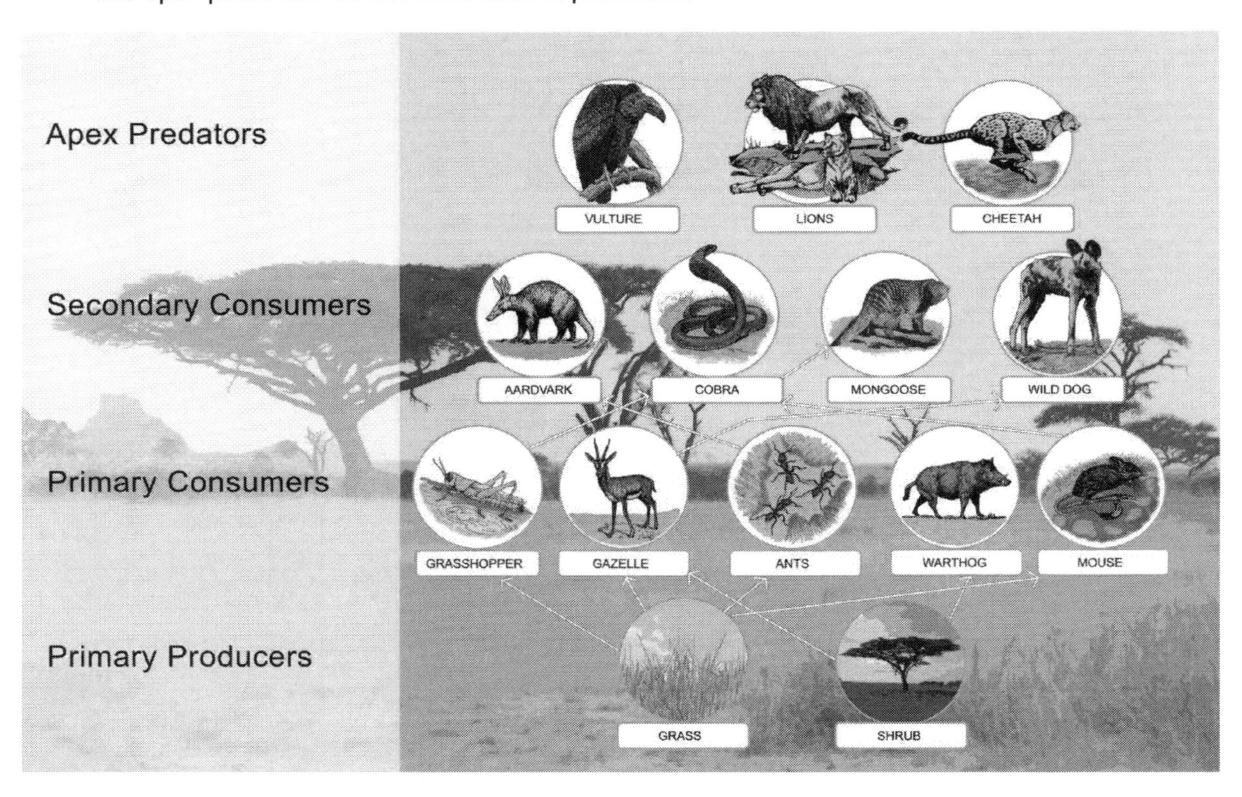

20. Which of the following eats primary producers according to the food chain diagram?
    a. Cobra
    b. Gazelle
    c. Wild dog
    d. Aardvark
    e. Grass

21. Which of the following animals has no natural predators according to the food chain diagram?
    a. Grass
    b. Cobra
    c. Mongoose
    d. Aardvark
    e. Vulture

22. Which of the following is something that the mongoose would eat?
    a. Shrub
    b. Aardvark
    c. Vulture
    d. Mouse
    e. Lion

Questions 23–26 are based on the timeline of the life of Alexander Graham Bell.

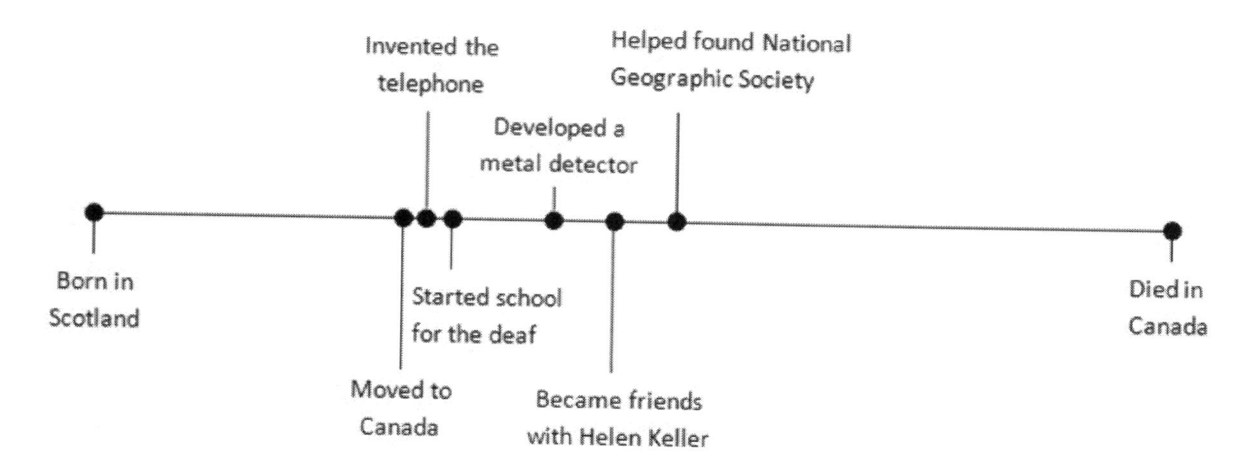

23. Which of the following is the event that occurred fourth on the timeline?
    a. Helped found National Geographic Society
    b. Developed a metal detector
    c. Moved to Canada
    d. Became friends with Helen Keller
    e. Started a school for the deaf

24. Of the pairings in the answer choices, which has the longest gap between the two events?
    a. Moved to Canada and Became friends with Helen Keller
    b. Became friends with Helen Keller and Died in Canada
    c. Started school for the deaf and Developed a metal detector
    d. Born in Scotland and Started school for the deaf
    e. Developed a metal detector and Became friends with Helen Keller

25. Which one of the following statements is accurate based on the timeline?
    a. Bell did nothing significant after he helped found the National Geographic Society.
    b. Bell started a school for the deaf in Canada.
    c. Bell lived in at least two countries.
    d. Developing a metal detector allowed Bell to meet Helen Keller.
    e. Bell was born and died in the same place.

26. Which one of the following events occurred most recently?
    a. Bell's invention of the telephone
    b. Bell's founding of the school
    c. Bell's birth
    d. Bell's move to Canada
    e. Bell's founding the National Geographic Society

Questions 27 and 28 are based on the following graph of high school students.

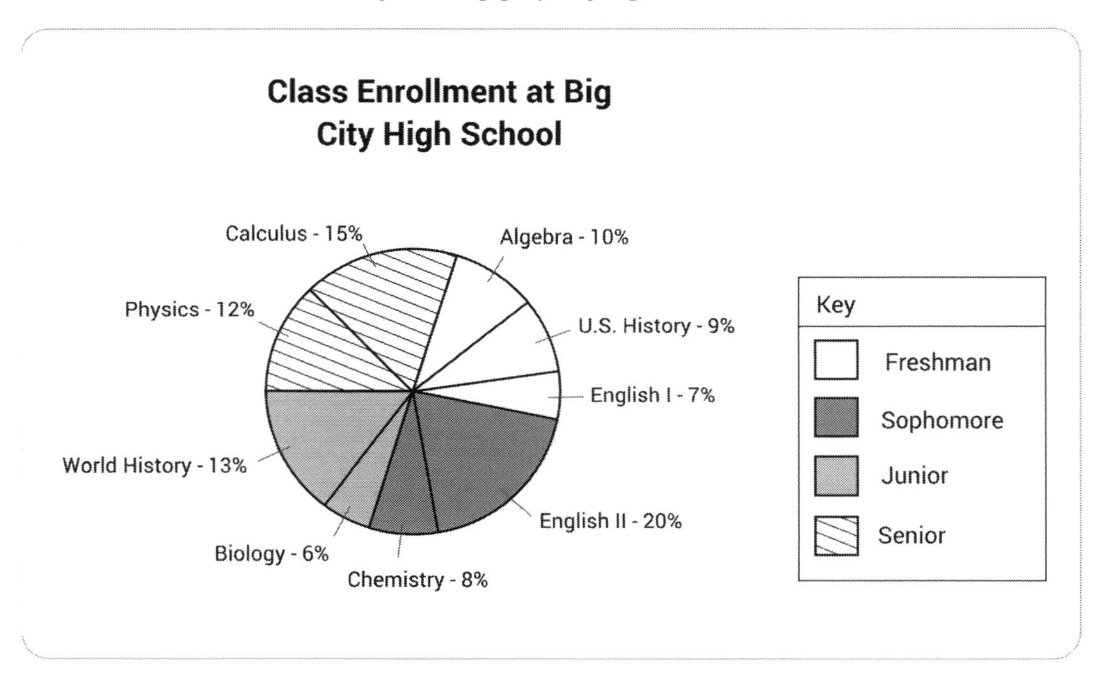

27. Which grade level has the most students in it?
    a. Freshman
    b. Sophomore
    c. Junior
    d. Senior
    e. It cannot be determined from the graph

28. What percent of the school is made up of freshman?
    a. 26%
    b. 27%
    c. 28%
    d. 29%
    e. 31

Questions 29–34 are based on the following passage:

> When researchers and engineers undertake a large-scale scientific project, they may end up making discoveries and developing technologies that have far wider uses than originally intended. This is especially true in NASA, one of the most influential and innovative scientific organizations in America. NASA *spinoff technology* refers to innovations originally developed for NASA space projects that are now used in a wide range of different commercial fields. Many

consumers are unaware that products they are buying are based on NASA research! Spinoff technology proves that it is worthwhile to invest in science research because it could enrich people's lives in unexpected ways.

The first spinoff technology worth mentioning is baby food. In space, where astronauts have limited access to fresh food and fewer options about their daily meals, malnutrition is a serious concern. Consequently, NASA researchers were looking for ways to enhance the nutritional value of astronauts' food. Scientists found that a certain type of algae could be added to food, improving the food's neurological benefits. When experts in the commercial food industry learned of this algae's potential to boost brain health, they were quick to begin their own research. The nutritional substance from algae then developed into a product called life's DHA, which can be found in over 90% of infant food sold in America.

Another intriguing example of a spinoff technology can be found in fashion. People who are always dropping their sunglasses may have invested in a pair of sunglasses with scratch resistant lenses—that is, it's impossible to scratch the glass, even if the glasses are dropped on an abrasive surface. This innovation is incredibly advantageous for people who are clumsy, but most shoppers don't know that this technology was originally developed by NASA. Scientists first created scratch resistant glass to help protect costly and crucial equipment from getting scratched in space, especially the helmet visors in space suits. However, sunglasses companies later realized that this technology could be profitable for their products, and they licensed the technology from NASA.

29. What is the main purpose of this article?
    a. To advise consumers to do more research before making a purchase
    b. To persuade readers to support NASA research
    c. To tell a narrative about the history of space technology
    d. To define and describe instances of spinoff technology
    e. To entertain readers with an engaging narrative

30. What is the organizational structure of this article?
    a. A general definition followed by more specific examples
    b. A general opinion followed by supporting arguments
    c. An important moment in history followed by chronological details
    d. A popular misconception followed by counterevidence
    e. An interweaving of comparisons and contrasts

31. Why did NASA scientists research algae?
    a. They already knew algae was healthy for babies.
    b. They were interested in how to grow food in space.
    c. They were looking for ways to add health benefits to food.
    d. They hoped to use it to protect expensive research equipment.
    e. They wanted to find a good use for the excessive algae near the laboratory.

32. What does the word "neurological" mean in the second paragraph?
    a. Related to the body
    b. Related to the brain
    c. Related to vitamins
    d. Related to technology

e. Related to birth

33. Why does the author mention space suit helmets?
    a. To give an example of astronaut fashion
    b. To explain where sunglasses got their shape
    c. To explain how astronauts protect their eyes
    d. To give an example of valuable space equipment
    e. Because people on Earth need helmets as well

34. Which statement would the author probably NOT agree with?
    a. Consumers don't always know the history of the products they are buying.
    b. Sometimes new innovations have unexpected applications.
    c. It is difficult to make money from scientific research.
    d. Space equipment is often very expensive.
    e. NASA has made more of an impact than simply exploring space.

35. Read the following poem. Which option best expresses the symbolic meaning of the "road" and the overall theme?

> Two roads diverged in a yellow wood,
> And sorry I could not travel both
> And be one traveler, long I stood
> And looked down one as far as I could
> To where it bent in the undergrowth;
> Then took the other, as just as fair,
> And having perhaps the better claim,
> Because it was grassy and wanted wear;
> Though as for that the passing there
> Had worn them really about the same,
> And both that morning equally lay
> In leaves no step had trodden black.
> Oh, I kept the first for another day!
> Yet knowing how way leads on to way,
> I doubted if I should ever come back.
> I shall be telling this with a sigh
> Somewhere ages and ages hence:
> Two roads diverged in a wood, and I—
> I took the one less traveled by,
> And that has made all the difference—Robert Frost, "The Road Not Taken"

    a. A divergent spot where the traveler had to choose the correct path to his destination
    b. A choice between good and evil that the traveler needs to make
    c. The traveler's struggle between his lost love and his future prospects
    d. Life's journey and the choices with which humans are faced
    e. The contemplation of death and possibilities for the afterlife.

36. Kimmy is a world-famous actress. Millions of people downloaded her leaked movie co-starring her previous boyfriend. Kimmy earns millions through her television show and marketing appearances. There's little wonder that paparazzi track her every move.

What is the argument's primary purpose?

a. Kimmy does not deserve her fame.
b. Kimmy starred in an extremely popular movie.
c. Kimmy earns millions of dollars through her television show and marketing appearances.
d. Kimmy is a highly compensated and extremely popular television and movie actress.
e. Kimmy has a huge family and a famous boyfriend.

37. Dwight works at a mid-sized regional tech company. He approaches all tasks with unmatched enthusiasm and leads the company in annual sales. The top salesman is always the best employee. Therefore, Dwight is the best employee.

Which of the following most accurately describes how the argument proceeds?

a. The argument proceeds by first stating a conclusion and then offering several premises to justify that conclusion.
b. The argument proceeds by stating a universal rule and then proceeds to show how this situation is the exception.
c. The argument proceeds by stating several facts that serve as the basis for the conclusion at the end of the argument.
d. The argument proceeds by stating several facts, offering a universal rule, and then drawing a conclusion by applying the facts to the rule.
e. The argument begins with an opinion and ends with a fact.

Questions 38–40 are based upon the following passage:

This excerpt is adaptation from "The 'Hatchery' of the Sun-Fish"--- *Scientific American,* #711

> I have thought that an example of the intelligence (instinct?) of a class of fish which has come under my observation during my excursions into the Adirondack region of New York State might possibly be of interest to your readers, especially as I am not aware that any one except myself has noticed it, or, at least, has given it publicity.
>
> The female sun-fish (called, I believe, in England, the roach or bream) makes a "hatchery" for her eggs in this wise. Selecting a spot near the banks of the numerous lakes in which this region abounds, and where the water is about 4 inches deep, and still, she builds, with her tail and snout, a circular embankment 3 inches in height and 2 thick. The circle, which is as perfect a one as could be formed with mathematical instruments, is usually a foot and a half in diameter; and at one side of this circular wall an opening is left by the fish of just sufficient width to admit her body.
>
> The mother sun-fish, having now built or provided her "hatchery," deposits her spawn within the circular inclosure, and mounts guard at the entrance until the fry are hatched out and are sufficiently large to take charge of themselves. As the embankment, moreover, is built up to the surface of the water, no enemy can very easily obtain an entrance within the inclosure from the top; while there being only one entrance, the fish is able, with comparative ease, to keep out all intruders.

> I have, as I say, noticed this beautiful instinct of the sun-fish for the perpetuity of her species more particularly in the lakes of this region; but doubtless the same habit is common to these fish in other waters.

38. What is the purpose of this passage?
    a. To show the effects of fish hatcheries on the Adirondack region
    b. To persuade the audience to study Ichthyology (fish science)
    c. To depict the sequence of mating among sun-fish
    d. To show the damaging effects of climate change on fish habitats.
    e. To enlighten the audience on the habits of sun-fish and their hatcheries

39. What does the word *wise* in this passage most closely mean?
    a. Knowledge
    b. Manner
    c. Shrewd
    d. Ignorance
    e. Intelligence

40. What is the definition of the word *fry* as it appears in the following passage?

    > The mother sun-fish, having now built or provided her "hatchery," deposits her spawn within the circular inclosure, and mounts guard at the entrance until the fry are hatched out and are sufficiently large to take charge of themselves.

    a. Fish at the stage of development where they are capable of feeding themselves.
    b. Fish eggs that have been fertilized.
    c. A place where larvae is kept out of danger from other predators.
    d. A dish where fish is placed in oil and fried until golden brown.
    e. A place in the stream where the mother lays her eggs.

Questions 41–46 are based on the following passage.

> Dana Gioia argues in his article that poetry is dying, now little more than a limited art form confined to academic and college settings. Of course, poetry remains healthy in the academic setting, but the idea of poetry being limited to this academic subculture is a stretch. New technology and social networking alone have contributed to poets and other writers' work being shared across the world. YouTube has emerged to be a major asset to poets, allowing live performances to be streamed to billions of users. Even now, poetry continues to grow and voice topics that are relevant to the culture of our time. Poetry is not in the spotlight as it may have been in earlier times, but it's still a relevant art form that continues to expand in scope and appeal.
>
> Furthermore, Gioia's argument does not account for live performances of poetry. Not everyone has taken a poetry class or enrolled in university—but most everyone is online. The Internet is a perfect launching point to get all creative work out there. An example of this was the performance of Buddy Wakefield's *Hurling Crowbirds at Mockingbars*. Wakefield is a well-known poet who has published several collections of contemporary poetry. One of my favorite works by Wakefield is *Crowbirds*, specifically his performance at New York University in 2009. Although his reading was a campus

event, views of his performance online number in the thousands. His poetry attracted people outside of the university setting.

Naturally, the poem's popularity can be attributed both to Wakefield's performance and the quality of his writing. *Crowbirds* touches on themes of core human concepts such as faith, personal loss, and growth. These are not ideas that only poets or students of literature understand, but all human beings: "You acted like I was hurling crowbirds at mockingbars / and abandoned me for not making sense. / Evidently, I don't experience things as rationally as you do" (Wakefield 15-17). Wakefield weaves together a complex description of the perplexed and hurt emotions of the speaker undergoing a separation from a romantic interest. The line "You acted like I was hurling crowbirds at mockingbars" conjures up an image of someone confused, seemingly out of their mind . . . or in the case of the speaker, passionately trying to grasp at a relationship that is fading. The speaker is looking back and finding the words that described how he wasn't making sense. This poem is particularly human and gripping in its message, but the entire effect of the poem is enhanced through the physical performance.

At its core, poetry is about addressing issues/ideas in the world. Part of this is also addressing the perspectives that are exiguously considered. Although the platform may look different, poetry continues to have a steady audience due to the emotional connection the poet shares with the audience.

41. Which one of the following best explains how the passage is organized?
   a. The author begins with a long definition of the main topic, and then proceeds to prove how that definition has changed over the course of modernity.
   b. The author presents a puzzling phenomenon and uses the rest of the passage to showcase personal experiences in order to explain it.
   c. The author contrasts two different viewpoints, then builds a case showing preference for one over the other.
   d. The passage is an analysis of another theory in which the author has no stake in.

42. The author of the passage would likely agree most with which of the following?
   a. Buddy Wakefield is a genius and is considered at the forefront of modern poetry.
   b. Poetry is not irrelevant; it is an art form that adapts to the changing time while containing its core elements.
   c. Spoken word is the zenith of poetic forms and the premier style of poetry in this decade.
   d. Poetry is on the verge of vanishing from our cultural consciousness.

43. Which one of the following words, if substituted for the word *exiguously* in the last paragraph, would LEAST change the meaning of the sentence?
   a. Indolently
   b. Inaudibly
   c. Interminably
   d. Infrequently

44. Which of the following is most closely analogous to the author's opinion of Buddy Wakefield's performance in relation to modern poetry?

a. Someone's refusal to accept that the Higgs Boson will validate the Standard Model.

b. An individual's belief that soccer will lose popularity within the next fifty years.

c. A professor's opinion that poetry contains the language of the heart, while fiction contains the language of the mind.

d. A student's insistence that psychoanalysis is a subset of modern psychology.

45. What is the primary purpose of the passage?

a. To educate readers on the development of poetry and describe the historical implications of poetry in media.

b. To disprove Dana Gioia's stance that poetry is becoming irrelevant and is only appreciated in academia.

c. To inform readers of the brilliance of Buddy Wakefield and to introduce them to other poets that have influence in contemporary poetry.

d. To prove that Gioia's article does have some truth to it and to shed light on its relevance to modern poetry.

46. What is the author's main reason for including the quote in the passage?

a. The quote opens up opportunity to disprove Gioia's views.

b. To demonstrate that people are still writing poetry even if the medium has changed in current times.

c. To prove that poets still have an audience to write for even if the audience looks different than it did centuries ago.

d. The quote illustrates the complex themes poets continue to address, which still draws listeners and appreciation.

Questions 47–50 are based upon the following passage.

This excerpt is adaptation from Mineralogy --- Encyclopedia International*, Grolier*

Mineralogy is the science of minerals, which are the naturally occurring elements and compounds that make up the solid parts of the universe. Mineralogy is usually considered in terms of materials in the Earth, but meteorites provide samples of minerals from outside the Earth.

A mineral may be defined as a naturally occurring, homogeneous solid, inorganically formed, with a definite chemical composition and an ordered atomic arrangement. The qualification *naturally occurring* is essential because it is possible to reproduce most minerals in the laboratory. For example, evaporating a solution of sodium chloride produces crystal indistinguishable from those of the mineral halite, but such laboratory-produced crystals are not minerals.

A *homogeneous solid* is one consisting of a single kind of material that cannot be separated into simpler compounds by any physical method. The requirement that a mineral be solid eliminates gases and liquids from consideration. Thus ice is a mineral (a very common one, especially at high altitudes and latitudes) but water is not. Some mineralogists dispute this restriction and would consider both water and native mercury (also a liquid) as minerals.

The restriction of minerals to *inorganically formed* substances eliminates those homogenous solids produced by animals and plants. Thus the shell of an oyster and the pearl inside, though both consist of calcium carbonate indistinguishable chemically or physically from the mineral aragonite comma are not usually considered minerals.

The requirement of a *definite chemical composition* implies that a mineral is a chemical compound, and the composition of a chemical compound is readily expressed by a formula. Mineral formulas may be simple or complex, depending upon the number of elements present and the proportions in which they are combined.

Minerals are crystalline solids, and the presence of an *ordered atomic arrangement* is the criterion of the crystalline state. Under favorable conditions of formation the ordered atomic arrangement is expressed in the external crystal form. In fact, the presence of an ordered atomic arrangement and crystalline solids was deduced from the external regularity of crystals by a French mineralogist, Abbé R. Haüy, early in the 19th century.

47. According to the text, an object or substance must have all of the following criteria to be considered a mineral except for?

a. Be naturally occurring
b. Be a homogeneous solid
c. Be organically formed
d. Have a definite chemical composition
e. Be reproducible in the lab

48. What is the definition of the word "homogeneous" as it appears in the following passage?

"A homogeneous solid is one consisting of a single kind of material that cannot be separated into simpler compounds by any physical method."

a. Made of similar substances
b. Differing in some areas
c. Having a higher atomic mass
d. Lacking necessary properties
e. A hybrid between a solid and a liquid

49. The suffix -logy refers to?

a. The properties of
b. The chemical makeup of
c. The study of
d. The classification of
e. The methods of

50. The author included the counterargument in the following passage to achieve which following effect?

> The requirement that a mineral be solid eliminates gases and liquids from consideration. Thus, ice is a mineral (a very common one, especially at high altitudes and latitudes) but water is not. Some mineralogists dispute this restriction and would consider both water and native mercury (also a liquid) as minerals.

a. To complicate the subject matter
b. To express a bias
c. To point to the fact that there are differing opinions in the field of mineralogy concerning the characteristics necessary to determine whether a substance or material is a mineral
d. To create a new subsection of minerals
e. To strengthen his or her argument and the persuasiveness of the text

Mathematics

1. Which of the following numbers has the greatest value?
   a. 1.4378
   b. 1.07548
   c. 1.43592
   d. 0.89409
   e. 1.43688

2. The value of 6 x 12 is the same as:
   a. 2 x 4 x 4 x 2
   b. 7 x 4 x 3
   c. 6 x 6 x 3
   d. 3 x 3 x 4 x 2
   e. 3 x 4 x 6 x 2

3. This chart indicates how many sales of CDs, vinyl records, and MP3 downloads occurred over the last year. Approximately what percentage of the total sales was from CDs?

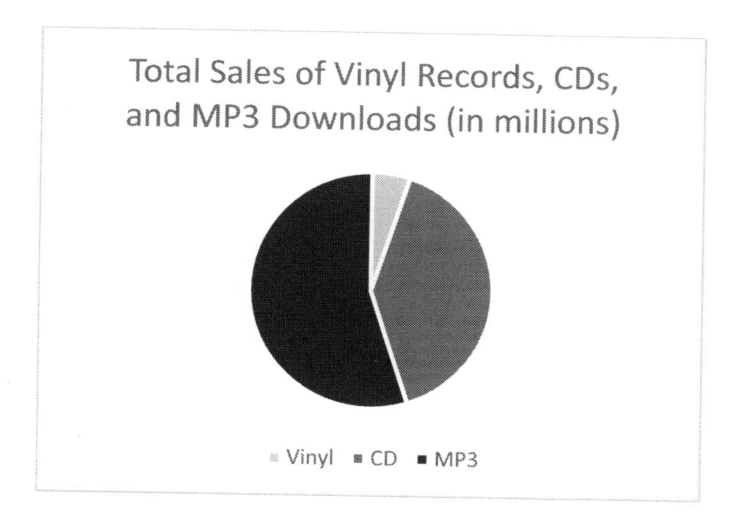

a. 55%
b. 25%
c. 40%
d. 5%
e. 20%

4. After a 20% sale discount, Frank purchased a new refrigerator for $850. How much did he save from the original price?
a. $170
b. $212.50
c. $105.75
d. $200
e. $150

5. A student gets an 85% on a test with 20 questions. How many answers did the student solve correctly?
a. 16
b. 15
c. 18
d. 19
e. 17

6. Alan currently weighs 200 pounds, but he wants to lose weight to get down to 175 pounds. What is this difference in kilograms? (1 pound is approximately equal to 0.45 kilograms.)
a. 9 kg
b. 11.25 kg
c. 78.75 kg
d. 90 kg
e. 25 kg

7. Johnny earns $2334.50 from his job each month. He pays $1437 for monthly expenses. Johnny is planning a vacation in 3 months' time that he estimates will cost $1750 total. How much will Johnny have left over from three months' of saving once he pays for his vacation?

a. $948.50
b. $584.50
c. $852.50
d. $942.50
e. $952.50

8. What is $\frac{420}{98}$ rounded to the nearest integer?

a. 3
b. 4
c. 5
d. 6
e.7

9. Dwayne has received the following scores on his math tests: 78, 92, 83, 97. What score must Dwayne get on his next math test to have an overall average of 90?

a. 89
b. 98
c. 95
d. 100
e. 96

10. What is the overall median of Dwayne's current scores: 78, 92, 83, 97?

a. 19
b. 85
c. 83
d. 87.5
e. 86

11. Solve the following:

$$(\sqrt{36} \times \sqrt{16}) - 3^2$$

a. 30
b. 21
c. 15
d. 13
e. 16

12. In Jim's school, there are 3 girls for every 2 boys. There are 650 students in total. Using this information, how many students are girls?

a. 260
b. 130
c. 65
d. 390
e. 225

13. Five of six numbers have a sum of 25. The average of all six numbers is 6. What is the sixth number?
a. 8
b. 12
c. 13
d. 10
e. 11

14. Kimberley earns $10 an hour babysitting, and after 10 p.m., she earns $12 an hour, with the amount paid being rounded to the nearest hour accordingly. On her last job, she worked from 5:30 p.m. to 11 p.m. In total, how much did Kimberley earn on her last job?
a. $45
b. $57
c. $62
d. $42
e. $67

15. Arrange the following numbers from least to greatest value:

$0.85, \frac{4}{5}, \frac{2}{3}, \frac{91}{100}$

a. $0.85, \frac{4}{5}, \frac{2}{3}, \frac{91}{100}$
b. $\frac{4}{5}, 0.85, \frac{91}{100}, \frac{2}{3}$
c. $\frac{2}{3}, \frac{4}{5}, 0.85, \frac{91}{100}$
d. $0.85, \frac{91}{100}, \frac{4}{5}, \frac{2}{3}$
e. $\frac{4}{5}, \frac{2}{3}, 0.85, \frac{91}{100}$

16. Keith's bakery had 252 customers go through its doors last week. This week, that number increased to 378. Express this increase as a percentage.
a. 26%
b. 50%
c. 35%
d. 12%
e. 28%

17. The following graph compares the various test scores of the top three students in each of these teacher's classes. Based on the graph, which teacher's students had the lowest range of test scores?

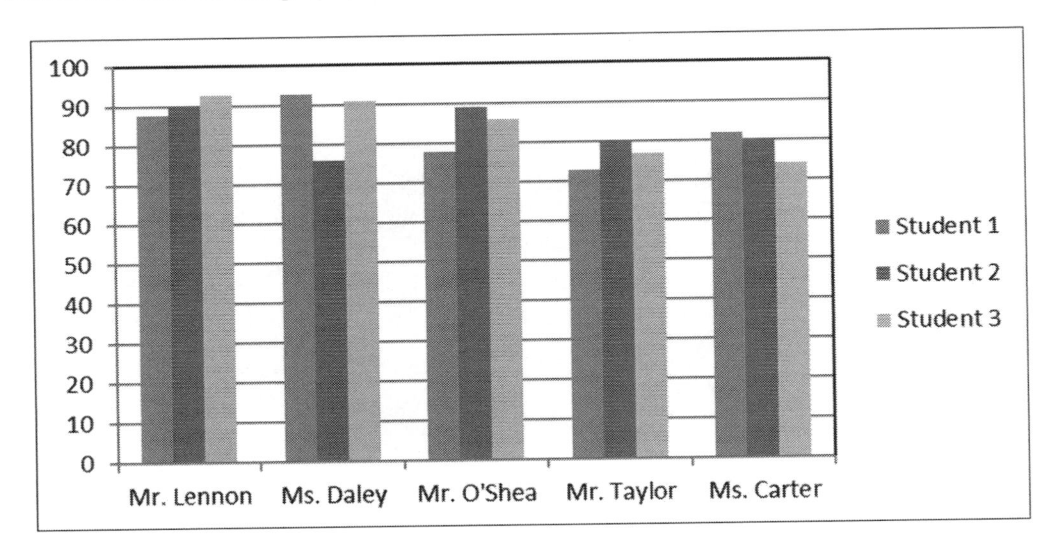

a. Mr. Lennon
b. Mr. O'Shea
c. Mr. Taylor
d. Ms. Daley
e. Ms. Carter

18. Four people split a bill. The first person pays for $\frac{1}{5}$, the second person pays for $\frac{1}{4}$, and the third person pays for $\frac{1}{3}$. What fraction of the bill does the fourth person pay?

a. $\frac{13}{60}$

b. $\frac{47}{60}$

c. $\frac{1}{4}$

d. $\frac{4}{15}$

e. $\frac{1}{2}$

19. Simplify the following expression:

$$4\frac{2}{3} - 3\frac{4}{9}$$

a. $1\frac{1}{3}$

b. $1\frac{2}{9}$

c. 1

d. $1\frac{2}{3}$

e. $1\frac{4}{9}$

20. A closet is filled with red, blue, and green shirts. If $\frac{1}{3}$ of the shirts are green and $\frac{2}{5}$ are red, what fraction of the shirts are blue?

a. $\frac{4}{15}$

b. $\frac{1}{5}$

c. $\frac{7}{15}$

d. $\frac{1}{2}$

e. $\frac{2}{3}$

21. Shawna buys $2\frac{1}{2}$ gallons of paint. If she uses $\frac{1}{3}$ of it on the first day, how much does she have left?

a. $1\frac{5}{6}$ gallons

b. $1\frac{1}{2}$ gallons

c. $1\frac{2}{3}$gallons

d. 2 gallons

e. $1\frac{3}{4}$ gallons

22. Jessica buys 10 cans of paint. Red paint costs $1 per can and blue paint costs $2 per can. In total, she spends $16. How many red cans did she buy?

a. 2
b. 3
c. 4
d. 5
e. 6

23. Six people apply to work for Janice's company, but she only needs four workers. How many different groups of four employees can Janice choose?

a. 6
b. 10
c. 15
d. 36
e. 30

24. Which of the following is equivalent to the value of the digit 3 in the number 792.134?

a. 3×10

b. 3×100

c. $\frac{3}{10}$

d. $\frac{3}{100}$

e. 3×0.1

25. In the following expression, which operation should be completed first? $5 \times 6 + (5 + 4) \div 2 - 1$.

a. Multiplication
b. Addition
c. Division
d. Subtraction
e. Parentheses

26. How will the number 847.89632 be written if rounded to the nearest hundredth?

a. 847.90
b. 900
c. 847.89
d. 847.896
e. 847.895

27. The perimeter of a 6-sided polygon is 56 cm. The length of three of the sides are 9 cm each. The length of two other sides are 8 cm each. What is the length of the missing side?

a. 11 cm
b. 12 cm
c. 13 cm
d. 10 cm
e. 9 cm

28. Which of the following is a mixed number?

a. $16\frac{1}{2}$

b. 16

c. $\frac{16}{3}$

d. $\frac{1}{4}$

e. $\frac{3}{6}$

29. Change 9.3 to a fraction.

a. $9\frac{3}{7}$

b. $\frac{903}{1000}$

c. $\frac{9.03}{100}$

d. $9\frac{3}{100}$

30. What is the value of b in this equation?

$5b - 4 = 2b + 17$

a. 13
b. 24
c. 7
d. 21
e. 14

31. Express the solution to the following problem in decimal form:

$$\frac{3}{5} \times \frac{7}{10} \div \frac{1}{2}$$

a. 0.042
b. 84%
c. 0.84
d. 0.42
e. 0.084

32. If $\frac{5}{2} \div \frac{1}{3} = n$, then n is between:

a. 5 and 7
b. 7 and 9
c. 9 and 11
d. 3 and 5
e. 13 and 15

33. Katie works at a clothing company and sold 192 shirts over the weekend. $\frac{1}{3}$ of the shirts that were sold were patterned, and the rest were solid. Which mathematical expression would calculate the number of solid shirts Katie sold over the weekend?

a. $192 \times \frac{1}{3}$
b. $192 \div \frac{1}{3}$
c. $192 \times (1 - \frac{1}{3})$
d. $192 \div 3$
e. $192 \div \left(1 - \frac{1}{3}\right)$

34. Which four-sided shape is always a rectangle?

a. Rhombus
b. Square
c. Parallelogram
d. Quadrilateral
e. Trapezoid

35. A rectangle was formed out of pipe cleaner. Its length was $\frac{1}{2}$ ft, and its width was $\frac{11}{2}$ inches. What is its area in square inches?

a. $\frac{11}{4}$ inch^2
b. $\frac{11}{2}$ inch^2
c. 22 inches^2
d. 33 inches^2
e. 11 inches^2

36. How will $\frac{4}{5}$ be written as a percent?

a. 40 percent
b. 125 percent
c. 90 percent
d. 80 percent
e. 85 percent

37. If Danny takes 48 minutes to walk 3 miles, how long should it take him to walk 5 miles maintaining the same speed?

a. 32 min
b. 64 min
c. 80 min
d. 96 min
e. 78 min

38. A solution needs 5 ml of saline for every 8 ml of medicine given. How much saline is needed for 45 ml of medicine?

a. $\frac{225}{8}$ ml

b. 72 ml

c. 28 ml

d. $\frac{45}{8}$ ml

e. 25 ml

39. What unit of volume is used to describe the following 3-dimensional shape?

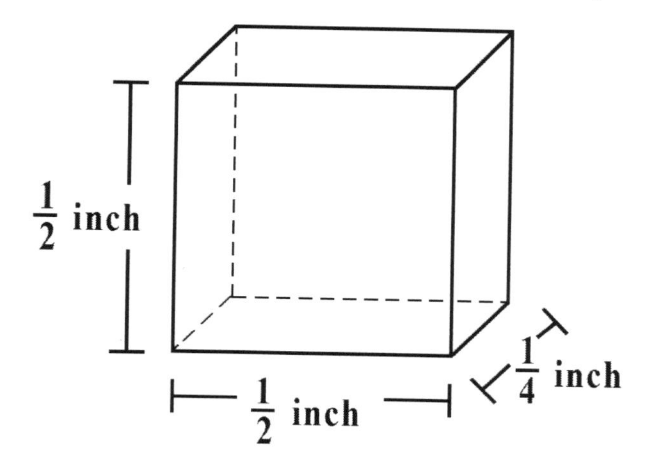

a. Square inches
b. Inches
c. Cubic inches
d. Squares
e. Square feet

40. Which common denominator would be used in order to evaluate $\frac{2}{3}+\frac{4}{5}$?
a. 15
b. 3
c. 5
d. 10
e. 8

41. The diameter of a circle measures 5.75 centimeters. What tool could be used to draw such a circle?
a. Ruler
b. Meter stick
c. Compass
d. Yard stick
e. Protractor

42. A piggy bank contains 12 dollars' worth of nickels. A nickel weighs 5 grams, and the empty piggy bank weighs 1050 grams. What is the total weight of the full piggy bank?

a. 1,110 grams
b. 1,200 grams
c. 2,150 grams
d. 2,200 grams
e. 2,250 grams

43. Last year, the New York City area received approximately $27\frac{3}{4}$ inches of snow. The Denver area received approximately 3 times as much snow as New York City. How much snow fell in Denver?

a. $71\frac{3}{4}$ inches
b. $27\frac{1}{4}$ inches
c. $89\frac{1}{4}$ inches
d. $83\frac{1}{4}$ inches
e. $86\frac{1}{2}$ inches

44. Which of the following would be an instance in which ordinal numbers are used?

a. Katie scored a 9 out of 10 on her quiz.
b. Matthew finished second in the spelling bee.
c. Jacob missed one day of school last month.
d. Kim was 5 minutes late to school this morning.
e. John was on vacation for 6 days.

45. The graph shows the position of a car over a 10-second time interval. Which of the following is the correct interpretation of the graph for the interval 1 to 3 seconds?

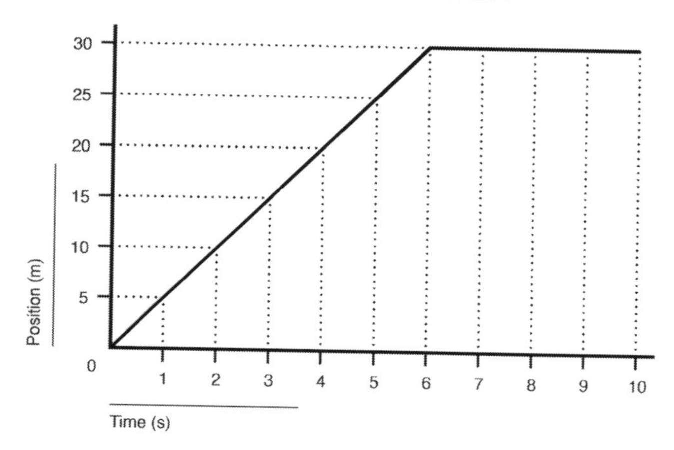

a. The car remains in the same position.
b. The car is traveling at a speed of 5m/s.
c. The car is traveling up a hill.
d. The car is traveling at 5mph.
e. The car is traveling at a speed of 10m/s.

46. If Sarah reads at an average rate of 21 pages in four nights, how long will it take her to read 140 pages?
a. 6 nights
b. 26 nights
c. 8 nights
d. 27 nights
e. 7 nights

47. The phone bill is calculated each month using the equation $c = 50g + 75$. The cost of the phone bill per month is represented by c, and g represents the gigabytes of data used that month. What is the value and interpretation of the slope of this equation?
a. 75 dollars per day
b. 75 gigabytes per day
c. 50 dollars per day
d. 50 dollars per gigabyte
e. 125 dollars per gigabyte

48. How will the following number be written in standard form:

$$(1 \times 10^4) + (3 \times 10^3) + (7 \times 10^1) + (8 \times 10^0)$$

a. 137
b. 13,780
c. 1,378
d. 8,731
e. 13,078

49. What is the area of the regular hexagon shown below?

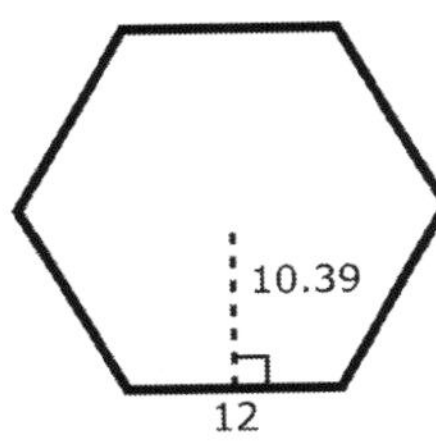

a. 72
b. 124.68
c. 374.04
d. 748.08
e. 372.96

50. The area of a given rectangle is 24 centimeters. If the measure of each side is multiplied by 3, what is the area of the new figure?

a. 48 cm^2
b. 72 cm^2
c. 216 cm^2
d. 13,824 cm^2
e. 224 cm^2

Writing

First Essay Prompt

Directions: The Writing section of the CBEST will require test takers to write two essays based on provided prompts. The first essay has a referential aim so that test takers can showcase their analytical and expository writing skills. The second essay will have an expressive aim, with a topic relating to the test taker's past lived experience.

Please read the prompt below and answer in an essay format.

> Some people feel that sharing their lives on social media sites such as Facebook, Instagram, and Snapchat is fine. They share every aspect of their lives, including pictures of themselves and their families, what they ate for lunch, who they are dating, and when they are going on vacation. They even say that if it's not on social media, it didn't happen. Other people believe that sharing so much personal information is an invasion of privacy and could prove dangerous. They think sharing personal pictures and details invites predators, cyberbullying, and identity theft.
>
> Write an essay to someone who is considering whether to participate in social media. Take a side on the issue and argue whether or not he/she should join a social media network. Use specific examples to support your argument.

Second Essay Prompt

Please read the prompt below and answer in an essay format.

> Students encounter many different courses and teachers throughout their high school and post-secondary careers. Pick a course that stands out in your memory and describe the influential impact and the elements or reasons that this course made a lasting impression on you.

Answer Explanations #3

Reading

1. B: Narrative, Choice *A*, means a written account of connected events. Think of narrative writing as a story. Choice C, expository writing, generally seeks to explain or describe some phenomena, whereas Choice *D*, technical writing, includes directions, instructions, and/or explanations. Choice *E,* informative writing, may be an appealing choice since this passage is informative in some ways, but informative texts should be very objective in their language. In contrast, this passage is definitely persuasive writing, which hopes to change someone's beliefs based on an appeal to reason or emotion. The author is aiming to convince the reader that smoking is terrible. They use health, price, and beauty in their argument against smoking, so Choice *B*, persuasive, is the correct answer. Persuasive is another term for argumentative.

2. B: The author is clearly opposed to tobacco. He cites disease and deaths associated with smoking. He points to the monetary expense and aesthetic costs. Choices *A* and *C* are wrong because they do not summarize the passage but rather are each just a premise. Choice *D* is wrong because, while these statistics are a premise in the argument, they do not represent a summary of the piece. Choice *B* is the correct answer because it states the three critiques offered against tobacco and expresses the author's conclusion. Choice *E* is wrong because alternatives to smoking are not even addressed in the passage.

3. C: We are looking for something the author would agree with, so it will almost certainly be anti-smoking or an argument in favor of quitting smoking. Choice *A* is wrong because the author does not speak against means of cessation. Choice *B* is wrong because the author does not reference other substances but does speak of how addictive nicotine, a drug in tobacco, is. Choice *D* is wrong because the author certainly would not encourage reducing taxes to encourage a reduction of smoking costs, thereby helping smokers to continue the habit. Choice *E* is wrong because the author states that according to the National Institute of Drug Abuse, nearly 35 million smokers expressed a desire to quit smoking in 2015. If the author had used the word "only" instead of "nearly" (and perhaps if the number was a lot lower) that would have changed the argument. Choice *C* is correct because the author is definitely attempting to persuade smokers to quit smoking.

4. D: Here, we are looking for an opinion of the author's rather than a fact or statistic. Choice *A* is wrong because quoting statistics from the Centers of Disease Control and Prevention is stating facts, not opinions. Choice *B* is wrong because it expresses the fact that cigarettes sometimes cost more than a few gallons of gas. It would be an opinion if the author said that cigarettes were not affordable. Choice *C* is incorrect because yellow stains are a known possible adverse effect of smoking. Choice *E* is incorrect because decreased life expectancy for smokers is a known fact because of the health problems it can create. Choice *D* is correct as an opinion because smell is subjective. Some people might like the smell of smoke, they might not have working olfactory senses, and/or some people might not find the smell of smoke akin to "pervasive nastiness," so this is the expression of an opinion. Thus, Choice *D* is the correct answer.

5. D: Although Washington was from a wealthy background, the passage does not say that his wealth led to his republican ideals, so Choice *A* is not supported. Choice *B* also does not follow from the passage. Washington's warning against meddling in foreign affairs does not mean that he would oppose wars of every kind, so Choice *B* is wrong. Choice *C* is also unjustified since the author does not indicate that Alexander Hamilton's assistance was absolutely necessary. Choice *E* is incorrect because we don't know which particular presidents Washington would befriend. Choice *D* is correct because the farewell address clearly opposes political parties and partisanship. The author then notes that presidential elections often hit a fever pitch of partisanship. Thus, it follows that George Washington would not approve of modern political parties and their involvement in presidential elections.

6. E: The author finishes the passage by applying Washington's farewell address to modern politics, so the purpose probably includes this application. Choice *A* does focus on modern politics, but it is incorrect. The passage does not state that it is the election process itself that George Washington would oppose; in fact, it mentions that he was also elected. Choice *B* is wrong because George Washington is already a well-established historical figure; furthermore, the passage does not seek to introduce him. Choice *C* is wrong because the author is not fighting a common perception that Washington was merely a military hero. Choice *D* is wrong because the author is not convincing readers. Persuasion does not correspond to the passage. Choice *E* states the primary purpose.

7. D: Choices *A* and *E* are wrong because the last paragraph is not appropriate for a history textbook nor a research journal, both which should be strictly objective and research-based. Choice *B* is false because the piece is not a notice or announcement of Washington's death. Choice *C* is clearly false because it is not fiction, but a historical writing. Choice *D* is correct. The passage is most likely to appear in a newspaper editorial because it cites information relevant and applicable to the present day, a popular format in editorials.

8. D: The passage does not proceed in chronological order since it begins by pointing out Leif Erikson's explorations in America so Choice *A* does not work. Although the author compares and contrasts Erikson with Christopher Columbus, this is not the main way the information is presented; therefore, Choice *B* does not work. Neither does Choice *C* because there is no mention of or reference to cause and effect in the passage. However, the passage does offer a conclusion (Leif Erikson deserves more credit) and premises (first European to set foot in the New World and first to contact the natives) to substantiate Erikson's historical importance. Thus, Choice *D* is correct. Choice *E* is incorrect because spatial order refers to the space and location of something or where things are located in relation to each other.

9. C: Choice *A* is wrong because it describes facts: Leif Erikson was the son of Erik the Red and historians debate Leif's date of birth. These are not opinions. Choice *B* is wrong; that Erikson called the land Vinland is a verifiable fact as is Choice *D* because he did contact the natives almost 500 years before Columbus. Choice *E* is also a fact and the passage mentions that there are several secondhand accounts (evidence) of their meetings. Choice *C* is the correct answer because it is the author's opinion that Erikson deserves more credit. That, in fact, is his conclusion in the piece, but another person could argue that Columbus or another explorer deserves more credit for opening up the New World to exploration. Rather than being an incontrovertible fact, it is a subjective value claim.

10. B: Choice *A* is wrong because the author aims to go beyond describing Erikson as a mere legendary Viking. Choice *C* is wrong because the author does not focus on Erikson's motivations, let alone name the spreading of Christianity as his primary objective. Choice *D* is wrong because it is a premise that Erikson contacted the natives 500 years before Columbus, which is simply a part of supporting the author's conclusion. Choice *E* is incorrect because the author states at the beginning that he or she believes it can't be considered "discovering" if people already lived there. Choice *B* is correct because, as stated in the previous answer, it accurately identifies the author's statement that Erikson deserves more credit than he has received for being the first European to explore the New World.

11. B: Choice *A* is wrong because the author is not in any way trying to entertain the reader. Choice *D* is wrong because he goes beyond a mere suggestion; "suggest" is too vague. Choice *E* is wrong for the same reason. Although the author is certainly trying to alert the readers of Leif Erikson's unheralded accomplishments, the nature of the writing does not indicate the author would be satisfied with the reader merely knowing of Erikson's exploration (Choice *C*). Rather, the author would want the reader to be informed about it, which is more substantial (Choice *B*).

12. D: Choice *A* is wrong because the author never addresses the Vikings' state of mind or emotions. Choice *B* is wrong because the author does not elaborate on Erikson's exile and whether he would have become an explorer if not for his banishment. Choice *C* is wrong because there is not enough information to support this premise. It is unclear whether Erikson informed the King of Norway of his finding. Although it is true that the King did not send a follow-up expedition, he could have simply chosen not to expend the resources after receiving Erikson's news. It is not possible to logically infer whether Erikson told him. Choice *E* is incorrect because the passage does not mention anything about Columbus' awareness of Erikson's travels. Choice *D* is correct because there are two examples—Leif Erikson's date of birth and what happened during the encounter with the natives—of historians having trouble pinning down important dates in Viking history.

13. E: The question asks which of the battles in the chart that are listed in this question had more Confederate casualties than Union casualties. There were more Confederate casualties than Union casualties at the Battles of Gettysburg and Atlanta. Of the two, only Atlanta is listed as an answer choice out of the battles explicitly listed or referred to by their dates. Thus, *E*, Atlanta, is the correct answer.

14. D: The question is asking you to find where the dates are located in the table and to identify the earliest battle. Answer D, Shiloh, occurred in April 1862 and no other battle listed on the table happened until May 1863. Choice *E,* Gettysburg, is listed first but did not occur first.

15. A: Robert E. Lee led the Confederate army in all battles listed on the table, except Shiloh and Atlanta. Looking at the dates and corresponding battles, as well as the battle explicitly listed, Shiloh is not listed as a choice, therefore Atlanta, option *A* is the correct answer.

16. A: This question is asking you to compare the Union and Confederate casualties and find the one listed where the Union casualties most exceeded the Confederate ones. At Cold Harbor, there were approximately 8,142 more Union casualties than Confederate. At Chancellorsville, there were approximately 3,844 more Union casualties. At Atlanta and Gettysburg, the number of Confederate casualties exceeded the Union number, so neither of these can be the correct answer. At Shiloh, there were approximately 2,378 more Union casualties. Thus, the number of Union casualties most greatly exceeding the Confederate number was at Cold Harbor, making *A*, Cold Harbor, the correct answer.

17. D: To calculate the total American casualties, we combine the Union and Confederate casualties since it was a civil war with Americans on both sides, so Choice *E* is incorrect. There were approximately 51,112 (Union + Confederate losses) American casualties at Gettysburg; thus, the correct answer will be a battle with approximately 25,556 casualties, which is half of that number. Shiloh is the closest with a total of 23,716 casualties, making *D* the correct answer.

18. E: The correct answer choice is *E,* 1910. There are two ways to arrive at the correct answer. You could find the five answer choices on the graph, or you could have identified that the population never dips at any point. Thus, the correct answer needs to be the only answer choice that is earlier in time than the others, Choice *E*.

19. D: The population increased the most between 1990 and 2010. The question is asking you to identify the rate of change for each interval. Between 1790 to 1810, the population increased by about 3.3 million. Between 1930 and 1950, the population increased by approximately 28 million. Between 1950 and 1970, the population increased by approximately 52 million. Between 1970 and 1990, the population increased by approximately 45 million. Between 1990 and 2010, the population increased by approximately 60 million. Thus, *D* is the correct answer. The slope is also the steepest in this interval, which represents its higher increase.

20. B: Primary producers make up the base of the food chain, so the correct answer will be in the level just above: a primary consumer. The cobra, wild dog, and aardvark are all secondary consumers. The gazelle is a primary consumer, so *B* is the correct answer. Grass *is* a primary producer, making *E* incorrect.

21. E: According to the passage preceding the food chain, the apex predators do not have natural predators. So, the question is really asking which of the answer choices is an apex predator. The cobra, mongoose, and aardvark are all secondary consumers. Grass is a primary producer. The vulture is an apex predator; thus, a vulture has no natural predators, making *E* the correct answer.

22. D: A mongoose is a secondary consumer; thus, the mongoose consumes primary consumers. The shrub is a primary producer. The aardvark is a secondary consumer. The vulture and the lion are apex predators. The mouse is a primary consumer, so Choice *D* is the correct answer.

23. E: This question is testing whether you realize how a timeline illustrates information in chronological order from left to right. "Started school for the deaf" is fourth on the timeline from the left, which means that it is the fourth event on the timeline. Thus, Choice *E* is the correct answer.

24. B: This question is asking you to determine the length of time between the pairs of events listed as answer choices. Events in timelines are arranged proportional to time. To determine the answer to this question, one must find the largest space between two events. Visually, this can be seen between the events of befriending Helen Keller and dying in Canada. Thus, Choice *B* is the correct answer.

25. C: This question is testing whether you can discern accurate conclusions from a timeline. Although the incorrect answer choices can seem correct, they cannot be confirmed from the information presented on the timeline. Choice *A* is incorrect; while it may be reasonable to assume that the timeline documents all major life events, we do not know for certain that Bell did not engage in any notable activities after founding the National Geographic Society. Choice *B* is incorrect because the timeline does not confirm that the school was in Canada; Bell actually started it in the United States. Choice *D* is incorrect because nothing on the timeline shows causation between the two events. Choice *E* is incorrect because the timeline states he was born in Scotland and died in Canada. Choice *C* is the only verifiable statement based on the timeline. Thus, *C* is the correct answer.

26. E: The founding of the National Geographic Society is the event listed farthest to the right of the events in the answer choices. This means it occurred most recently. Thus, *E* is the correct answer.

27. B: The sophomore class is made up of English II and Chemistry and encompasses 28% of the students. Freshman only make up 26% of the students, while Seniors make up 27%, and Juniors make up 19%.

28. A: To find this add up the percent of Algebra, US History, and English I. This comes up to 26%.

29. D: To define and describe instances of spinoff technology. This is an example of a purpose question—*why* did the author write this? The article contains facts, definitions, and other objective information without telling a story or arguing an opinion. In this case, the purpose of the article is to inform the reader. The only answer choice that is related to giving information is answer Choice *D*: to define and describe.

30. A: A general definition followed by more specific examples. This organization question asks readers to analyze the structure of the essay. The topic of the essay is about spinoff technology; the first paragraph gives a general definition of the concept, while the following two paragraphs offer more detailed examples to help illustrate this idea.

31. C: They were looking for ways to add health benefits to food. This reading comprehension question can be answered based on the second paragraph—scientists were concerned about astronauts' nutrition and began researching useful nutritional supplements. Choice *A* in particular is not true because it reverses the order of discovery (first NASA identified algae for astronaut use, and then it was further developed for use in baby food).

32. B: Related to the brain. This vocabulary question could be answered based on the reader's prior knowledge; but even for readers who have never encountered the word "neurological" before, the passage does provide context clues. The very next sentence talks about "this algae's potential to boost brain health," which is a paraphrase of "neurological benefits." From this context, readers should be able to infer that "neurological" is related to the brain.

33. D: To give an example of valuable space equipment. This purpose question requires readers to understand the relevance of the given detail. In this case, the author mentions "costly and crucial equipment" before mentioning space suit visors, which are given as an example of something that is very valuable. *A* is not correct because fashion is only related to sunglasses, not to NASA equipment. *B* can be eliminated because it is simply not mentioned in the passage. While *C* seems like it could be a true statement, it is also not relevant to what is being explained by the author. A similar thing could be said about *E*. The passage does not talk about helmet use on Earth.

34. C: It is difficult to make money from scientific research. The article gives several examples of how businesses have been able to capitalize on NASA research, so it is unlikely that the author would agree with this statement. Evidence for the other answer choices can be found in the article: *A*, the author mentions that "many consumers are unaware that products they are buying are based on NASA research"; *B* is a general definition of spinoff technology; *D* is mentioned in the final paragraph, and *E* is mentioned in the opening paragraph and is a theme throughout.

35. D: Choice *D* correctly summarizes Frost's theme of life's journey and the choices one makes. While Choice *A* can be seen as an interpretation, it is a literal one and is incorrect. Literal is not symbolic. Choice *B* presents the idea of good and evil as a theme, and the poem does not specify this struggle for the traveler. Choice *C* is a similarly incorrect answer. Love is not the theme. Choice *E* is incorrect because we don't receive any indication of death as the topic of this poem.

36. D: Choice *D* accurately expresses the argument's conclusion, and it best describes the argument's primary purpose. The argument concludes that Kimmy is a world-famous actress. Choice *D* is the best expression of the argument's purpose. Therefore, Choice *D* is the correct answer. Choice *A* is irrelevant. The argument does not address whether Kimmy deserves her fame. Eliminate this choice. Choice *B* restates a premise. Kimmy starring in an extremely popular movie is only one piece of the argument. It is not the main purpose. Eliminate this choice. Choice *C* also restates a premise, and it is incorrect for the same reasons as Choice *B*. Eliminate this choice. Choice *E* is incorrect; we are not told this information in the paragraph.

37. D: Choice *D* looks extremely promising. The argument first states several facts—Dwight works at a mid-sized regional tech company and leads the company in sales—then states a rule. Lastly, the argument applies the facts to the rule and concludes that Dwight is the best employee. This is a better fit than Choice *C* since it includes the rule and its application. Choice *A* is clearly incorrect. The argument does not start with a conclusion. Eliminate this choice. Choice *B* is incorrect. Although the argument states a universal rule—the top salesman is always a company's best employee—it does not argue that Dwight is the exception. Eliminate this choice. Choice *C* is fairly strong. The argument does state several facts and offers a conclusion based on those facts. Leave this choice for now. Therefore, Choice *D* is the correct answer. Choice *E* is not the best answer; the fact that Dwight works at a tech company is a fact, not an opinion.

38. E: To enlighten the audience on the habits of sun-fish and their hatcheries. Choice *A* is incorrect because although the Adirondack region is mentioned in the text, there is no cause or effect relationships between the region and fish hatcheries depicted here. Choice *B* is incorrect because the text does not have an agenda, but rather is meant to inform the audience. Choice *C* is incorrect because the text says nothing of how sun-fish mate. Choice *D* is incorrect; there is nothing mentioned about climate change in this passage.

39. B: The word *wise* in this passage most closely means *manner*. Choices *A, C,* and *E* are synonyms of *wise*; however, they are not relevant in the context of the text. Choice *D*, *ignorance*, is opposite of the word *wise*, and is therefore incorrect.

40. A: Fish at the stage of development where they are capable of feeding themselves. Even if the word *fry* isn't immediately known to the reader, the context gives a hint when it says "until the fry are hatched out and are sufficiently large to take charge of themselves."

41. C: The author contrasts two different viewpoints, then builds a case showing preference for one over the other. Choice *A* is incorrect because the introduction does not contain an impartial definition, but rather, an opinion. Choice *B* is incorrect. There is no puzzling phenomenon given, as the author doesn't mention any peculiar cause or effect that is in question regarding poetry. Choice *D* does contain another's viewpoint at the beginning of the passage; however, to say that the author has no stake in this argument is incorrect; the author uses personal experiences to build their case.

42. B: Choice *B* accurately describes the author's argument in the text: that poetry is not irrelevant. While the author does praise, and even value, Buddy Wakefield as a poet, the author never heralds him as a genius. Eliminate Choice *A*, as it is an exaggeration. Not only is Choice *C* an exaggerated statement, but the author never mentions spoken word poetry in the text. Choice *D* is wrong because this statement contradicts the writer's argument.

43. D: *Exiguously* means not occurring often, or occurring rarely, so Choice *D* would LEAST change the meaning of the sentence. Choice *A*, *indolently*, means unhurriedly, or slow, and does not fit the context of the sentence. Choice *B*, *inaudibly*, means quietly or silently. Choice *C*, *interminably*, means endlessly, or all the time, and is the opposite of the word *exiguously*.

44. D: A student's insistence that psychoanalysis is a subset of modern psychology is the most analogous option. The author of the passage tries to insist that performance poetry is a subset of modern poetry, and therefore, tries to prove that modern poetry is not "dying," but thriving on social media for the masses. Choice *A* is incorrect, as the author is not refusing any kind of validation. Choice *B* is incorrect; the author's insistence is that poetry will *not* lose popularity. Choice *C* mimics the topic but compares two different genres, while the author does no comparison in this passage.

45. B: The author's purpose is to disprove Gioia's article claiming that poetry is a dying art form that only survives in academic settings. In order to prove his argument, the author educates the reader about new developments in poetry (Choice *A*) and describes the brilliance of a specific modern poet (Choice *C*), but these serve as examples of a growing poetry trend that counters Gioia's argument. Choice *D* is incorrect because it contradicts the author's argument.

46. D: This question is difficult because the choices offer real reasons as to why the author includes the quote. However, the question specifically asks for the *main reason* for including the quote. The quote from a recently written poem shows that people are indeed writing, publishing, and performing poetry (Choice *B*). The quote also shows that people are still listening to poetry (Choice *C*). These things are true, and by their nature, serve to disprove Gioia's views (Choice *A*), which is the author's goal. However, Choice *D* is the most direct reason for including the quote, because the article analyzes the quote for its "complex themes" that "draws listeners and appreciation" right after it's given.

47. C: The text mentions all of the listed properties of minerals except the instance of minerals being organically formed. Objects or substances must be naturally occurring, must be a homogeneous solid, and must have a definite chemical composition in order to be considered a mineral.

48. A: Choice *A* is the correct answer because the prefix "homo" means same. Choice *B* is incorrect because "differing in some areas" would be linked to the root word "hetero," meaning "different" or "other."

49: C: Choice *C* is the correct answer because *-logy* refers to the study of a particular subject matter.

50: C: Choice *C* is the correct answer because the counterargument is necessary to point to the fact that researchers don't always agree with findings. Choices *A* and *B* are incorrect because the counterargument isn't overcomplicated or expressing bias, but simply stating an objective dispute. Choice *D* is incorrect because the counterargument is not used to persuade readers to create a new subsection of minerals.

Mathematics

1. A: Compare each numeral after the decimal point to figure out which overall number is greatest. In answers *A* (1.43785) and *C* (1.43592), both have the same tenths (4) and hundredths (3). However, the thousandths is greater in answer *A* (7), so *A* has the greatest value overall.

2. D: By grouping the four numbers in the answer into factors of the two numbers of the question (6 and 12), it can be determined that (3 x 2) x (4 x 3) = 6 x 12. Alternatively, each of the answer choices could be prime factored or multiplied out and compared to the original value. 6×12 has a value of 72 and a prime factorization of $2^3 \times 3^2$. The answer choices respectively have values of 64, 84, 108, 72, and 144 and prime factorizations of 2^6, $2^2 \times 3 \times 7$, $2^2 \times 3^3$, and $2^3 \times 3^2$, so answer *D* is the correct choice.

3. C: The sum total percentage of a pie chart must equal 100%. Since the CD sales take up less than half of the chart and more than a quarter (25%), it can be determined to be 40% overall. This can also be measured with a protractor. The angle of a circle is 360°. Since 25% of 360 would be 90° and 50% would be 180°, the angle percentage of CD sales falls in between; therefore, it would be answer *C*.

4. B: Since \$850 is the price *after* a 20% discount, \$850 represents 80% of the original price. To determine the original price, set up a proportion with the ratio of the sale price (850) to original price (unknown) equal to the ratio of sale percentage:

$$\frac{850}{x} = \frac{80}{100}$$

(where *x* represents the unknown original price)

To solve a proportion, cross multiply the numerators and denominators and set the products equal to each other: (850)(100) = (80)(x). Multiplying each side results in the equation 85,000=80x.

To solve for *x*, divide both sides by 80: $\frac{85{,}000}{80} = \frac{80x}{80}$, resulting in *x*=1062.5. Remember that *x* represents the original price. Subtracting the sale price from the original price (\$1062.50-\$850) indicates that Frank saved \$212.50.

5. E: 85% of a number means that number should be multiplied by 0.85: $0.85 \times 20 = \frac{85}{100} \times \frac{20}{1}$, which can be simplified to:

$$\frac{17}{20} \times \frac{20}{1} = 17$$

6. B: Using the conversion rate, multiply the projected weight loss of 25 lb by 0.45 $\frac{kg}{lb}$ to get the amount in kilograms (11.25 kg).

7. D: First, subtract \$1437 from \$2334.50 to find Johnny's monthly savings; this equals \$897.50. Then, multiply this amount by 3 to find out how much he will have (in three months) before he pays for his vacation: this equals \$2692.50. Finally, subtract the cost of the vacation (\$1750) from this amount to find how much Johnny will have left: \$942.50.

8. B: Dividing by 98 can be approximated by dividing by 100, which would mean shifting the decimal point of the numerator to the left by 2. The result is 4.2 which rounds to 4.

9. D: To find the average of a set of values, add the values together and then divide by the total number of values. In this case, include the unknown value of what Dwayne needs to score on his next test, in order to solve it.

$$\frac{78 + 92 + 83 + 97 + x}{5} = 90$$

Add the unknown value to the new average total, which is 5. Then multiply each side by 5 to simplify the equation, resulting in:

$78 + 92 + 83 + 97 + x = 450$
$350 + x = 450$
$x = 100$

Dwayne would need to get a perfect score of 100 in order to get an average of at least 90.

Test this answer by substituting back into the original formula.

$$\frac{78 + 92 + 83 + 97 + 100}{5} = 90$$

10. D: For an even number of total values, the *median* is calculated by finding the *mean* or average of the two middle values once all values have been arranged in ascending order from least to greatest. In this case, $(92 + 83) \div 2$ would equal the median 87.5, answer *D*.

11. C: Follow the *order of operations* in order to solve this problem. Solve the parentheses first, and then follow the remainder as usual.

$$(6 \times 4) - 9$$

This equals 24 – 9 or 15, answer *C*.

12. D: Three girls for every two boys can be expressed as a ratio: 3:2. This can be visualized as splitting the school into 5 groups: 3 girl groups and 2 boy groups. The number of students which are in each group can be found by dividing the total number of students by 5:

650 divided by 5 equals 1 part, or 130 students per group

To find the total number of girls, multiply the number of students per group (130) by the number of girl groups in the school (3). This equals 390, which is answer D.

13. E: If the average of all six numbers is 6, that means:

$$\frac{a + b + c + d + e + x}{6} = 6$$

The sum of the first five numbers is 25, so this equation can be simplified to $\frac{25+x}{6} = 6$. Multiplying both sides by 6 gives $25 + x = 36$, and *x*, or the sixth number, can be solved to equal 11.

14. C: Kimberley worked 4.5 hours at the rate of $10/h and 1 hour at the rate of $12/h. The problem states that her pay is rounded to the nearest hour, so the 4.5 hours would round up to 5 hours at the rate of $10/h.

$$(5h)\left(\frac{\$10}{h}\right) + (1h)\left(\frac{\$12}{h}\right) = \$50 + \$12 = \$62$$

15. C: The first step is to depict each number using decimals. $\frac{91}{100}$ = 0.91

Dividing the numerator by denominator of $\frac{4}{5}$ to convert it to a decimal yields 0.80, while $\frac{2}{3}$ becomes 0.66 recurring. Rearrange each expression in ascending order, as found in answer C.

16. B: First, calculate the difference between the larger value and the smaller value.

378 – 252 = 126

17. A: To calculate the range in a set of data, subtract the highest value with the lowest value. In this graph, the range of Mr. Lennon's students is 5, which can be seen physically in the graph as having the smallest difference compared with the other teachers between the highest value and the lowest value.

18. A: To find the fraction of the bill that the first three people pay, the fractions need to be added, which means finding common denominator. The common denominator will be 60.

$$\frac{1}{5}+\frac{1}{4}+\frac{1}{3}=\frac{12}{60}+\frac{15}{60}+\frac{20}{60}=\frac{47}{60}$$

The remainder of the bill is:

$$1-\frac{47}{60}=\frac{60}{60}-\frac{47}{60}=\frac{13}{60}$$

19. B: Simplify each mixed number of the problem into a fraction by multiplying the denominator by the whole number and adding the numerator:

$$\frac{14}{3}-\frac{31}{9}$$

Since the first denominator is a multiple of the second, simplify it further by multiplying both the numerator and denominator of the first expression by 3 so that the denominators of the fractions are equal.

$$\frac{42}{9}-\frac{31}{9}=\frac{11}{9}$$

Simplifying this further, divide the numerator 11 by the denominator 9; this leaves 1 with a remainder of 2. To write this as a mixed number, place the remainder over the denominator, resulting in $1\frac{2}{9}$.

20. A: The total fraction taken up by green and red shirts will be:

$$\frac{1}{3}+\frac{2}{5}=\frac{5}{15}+\frac{6}{15}=\frac{11}{15}$$

The remaining fraction is:

$$1-\frac{11}{15}=\frac{15}{15}-\frac{11}{15}=\frac{4}{15}$$

21. C: If she has used 1/3 of the paint, she has 2/3 remaining. $2\frac{1}{2}$ gallons are the same as $\frac{5}{2}$ gallons. The calculation is:

$$\frac{2}{3}\times\frac{5}{2}=\frac{5}{3}=1\frac{2}{3}\text{ gallons}$$

22. C: We are trying to find x, the number of red cans. The equation can be set up like this:

$$x + 2(10 - x) = 16$$

The left x is actually multiplied by $1, the price per red can. Since we know Jessica bought 10 total cans, $10 - x$ is the number blue cans that she bought. We multiply the number of blue cans by $2, the price per blue can.

That should all equal $16, the total amount of money that Jessica spent. Working that out gives us:

$$x + 20 - 2x = 16$$

$$20 - x = 16$$

$$x = 4$$

23. C: Janice will be choosing 4 employees out of a set of 6 applicants, so this will be given by the choice function. The following equation shows the choice function worked out:

$$\binom{6}{4} = \frac{6!}{4!\,(6-4)!} = \frac{6!}{4!\,(2)!}$$

$$\frac{6 \cdot 5 \cdot 4 \cdot 3 \cdot 2 \cdot 1}{4 \cdot 3 \cdot 2 \cdot 1 \cdot 2 \cdot 1} = \frac{6 \cdot 5}{2} = 15$$

24. D:$\frac{3}{100}$. Each digit to the left of the decimal point represents a higher multiple of 10 and each digit to the right of the decimal point represents a quotient of a higher multiple of 10 for the divisor. The first digit to the right of the decimal point is equal to the value $\div$ 10. The second digit to the right of the decimal point is equal to the value $\div$ (10×10), or the value $\div$ 100.

25. E: Using the order of operations, multiplication and division are computed first from left to right. Multiplication is on the left; therefore, the teacher should perform multiplication first.

26. A:847.90. The hundredth place value is located two digits to the right of the decimal point (the digit 9). The digit to the right of the place value is examined to decide whether to round up or keep the digit. In this case, the digit 6 is 5 or greater so the hundredth place is rounded up. When rounding up, if the digit to be increased is a 9, the digit to its left is increased by one and the digit in the desired place value is made a zero. Therefore, the number is rounded to 847.90.

27. C: Perimeter is found by calculating the sum of all sides of the polygon. $9 + 9 + 9 + 8 + 8 + s = 56$, where s is the missing side length. Therefore, 43 plus the missing side length is equal to 56. The missing side length is 13 cm.

28. A:$16\frac{1}{2}$. A mixed number contains both a whole number and either a fraction or a decimal. Therefore, the mixed number is $16\frac{1}{2}$.

29. D: $9\frac{3}{10}$

To convert a decimal to a fraction, remember that any number to the left of the decimal point will be a whole number. Then, since 0.3 goes to the tenths place, it can be placed over 10.

30. C: To solve for the value of b, both sides of the equation need to be equalized.

Start by cancelling out the lower value of -4 by adding 4 to both sides:

$5b - 4 = 2b + 17$
$5b - 4 + 4 = 2b + 17 + 4$
$5b = 2b + 21$

The variable *b* is the same on each side, so subtract the lower 2b from each side:

$5b = 2b + 21$
$5b - 2b = 2b + 21 - 2b$
$3b = 21$

Then divide both sides by 3 to get the value of *b*:

$$3b = 21$$

$$\frac{3b}{3} = \frac{21}{3}$$

$$b = 7$$

31. C: The first step in solving this problem is expressing the result in fraction form. Separate this problem first by solving the division operation of the last two fractions. When dividing one fraction by another, invert or flip the second fraction and then multiply the numerator and denominator.

$$\frac{7}{10} \times \frac{2}{1} = \frac{14}{10}$$

Next, multiply the first fraction with this value:

$$\frac{3}{5} \times \frac{14}{10} = \frac{42}{50}$$

Decimals are expressions of 1 or 100%, so multiply both the numerator and denominator by 2 to get the fraction as an expression of 100.

$$\frac{42}{50} \times \frac{2}{2} = \frac{84}{100}$$

In decimal form, this would be expressed as 0.84.

32. B: $\frac{5}{2} \div \frac{1}{3} = \frac{5}{2} \times \frac{3}{1} = \frac{15}{2} = 7.5$.

33. C: $\frac{1}{3}$ of the shirts sold were patterned. Therefore, $1 - \frac{1}{3} = \frac{2}{3}$ of the shirts sold were solid. Anytime "of" a quantity appears in a word problem, multiplication needs to be used. Therefore:

$$192 \times \frac{2}{3} = 192 \times \frac{2}{3} = \frac{384}{3} = 128 \text{ solid shirts were sold}$$

The entire expression is $192 \times \left(1 - \frac{1}{3}\right)$.

34. B: A rectangle is a specific type of parallelogram. It has 4 right angles. A square is a rhombus that has 4 right angles. Therefore, a square is always a rectangle because it has two sets of parallel lines and 4 right angles.

35. D: Area = length x width. The answer must be in square inches, so all values must be converted to inches. $\frac{1}{2}$ ft is equal to 6 inches. Therefore, the area of the rectangle is equal to:

$$6 \times \frac{11}{2} = \frac{66}{2} = 33 \text{ square inches}$$

36. D:80 percent. To convert a fraction to a percent, the fraction is first converted to a decimal. To do so, the numerator is divided by the denominator: $4 \div 5 = 0.8$. To convert a decimal to a percent, the number is multiplied by 100: $0.8 \times 10 = 80\%$.

37. C: 80 min. To solve the problem, a proportion is written consisting of ratios comparing distance and time. One way to set up the proportion is: $\frac{3}{48} = \frac{5}{x} \left(\frac{distance}{time} = \frac{distance}{time}\right)$ where *x* represents the unknown value of time. To solve a proportion, the ratios are cross-multiplied:

$$(3)(x) = (5)(48) \rightarrow 3x = 240$$

The equation is solved by isolating the variable, or dividing by 3 on both sides, to produce $x = 80$.

38. A: Every 8 ml of medicine requires 5 ml. The 45 ml first needs to be split into portions of 8 ml. This results in $\frac{45}{8}$portions. Each portion requires 5 ml. Therefore:

$$\frac{45}{8} \times 5 = 45 \times \frac{5}{8} = \frac{225}{8} \text{ ml is necessary}$$

39. C: Volume of this 3-dimensional figure is calculated using length x width x height. Each measure of length is in inches. Therefore, the answer would be labeled in cubic inches.

40. A: A common denominator must be found. The least common denominator is 15 because it has both 5 and 3 as factors. The fractions must be rewritten using 15 as the denominator.

41. C: A compass is a tool that can be used to draw a circle. The compass would be drawn by using the length of the radius, which is half of the diameter.

42. E: A dollar contains 20 nickels. Therefore, if there are 12 dollars' worth of nickels, there are $12 \times 20 = 240$ nickels. Each nickel weighs 5 grams. Therefore, the weight of the nickels is $240 \times 5 = 1{,}200$ grams. Adding in the weight of the empty piggy bank, the filled bank weighs 2,250 grams.

43. D: 3 must be multiplied times $27\frac{3}{4}$. In order to easily do this, the mixed number should be converted into an improper fraction.

$$27\frac{3}{4} = 27 \times 4 + \frac{3}{4} = \frac{111}{4}$$

Therefore, Denver had approximately:

$$3 \times \frac{111}{4} = \frac{333}{4} \text{ inches of snow}$$

The improper fraction can be converted back into a mixed number through division.

$$\frac{333}{4} = 83\frac{1}{4} \text{ inches}$$

44. B: Ordinal numbers represent a ranking. Placing second in a competition is a ranking among the other participants of the spelling bee.

45. B: The car is traveling at a speed of five meters per second. On the interval from one to three seconds, the position changes by fifteen meters. By making this change in position over time into a rate, the speed becomes ten meters in two seconds or five meters in one second.

46. D: This problem can be solved by setting up a proportion involving the given information and the unknown value. The proportion is:

$$\frac{21\,pages}{4\,nights} = \frac{140\,pages}{x\,nights}$$

Solving the proportion by cross-multiplying, the equation becomes $21x = 4 * 140$, where $x = 26.67$. Since it is not an exact number of nights, the answer is rounded up to 27 nights. Twenty-six nights would not give Sarah enough time.

47. D: The slope from this equation is 50, and it is interpreted as the cost per gigabyte used. Since the g-value represents number of gigabytes and the equation is set equal to the cost in dollars, the slope relates these two values. For every gigabyte used on the phone, the bill goes up 50 dollars.

48. E: 13,078. The power of 10 by which a digit is multiplied corresponds with the number of zeros following the digit when expressing its value in standard form. Therefore:

$$(1 \times 10^4) + (3 \times 10^3) + (7 \times 10^1) + (8 \times 10^0)$$

$$10{,}000 + 3{,}000 + 70 + 8 = 13{,}078$$

49. C: 374.04. The formula for finding the area of a regular polygon is $A = \frac{1}{2} \times a \times P$ where *a* is the length of the apothem (from the center to any side at a right angle), and P is the perimeter of the figure. The apothem *a* is given as 10.39, and the perimeter can be found by multiplying the length of one side by the number of sides (since the polygon is regular):

$$P = 12 \times 6 \rightarrow P = 72$$

To find the area, substitute the values for *a* and *P* into the formula:

$$A = \frac{1}{2} \times a \times P \rightarrow A = \frac{1}{2} \times (10.39) \times (72) \rightarrow A = 374.04$$

50. C: 216cm. Because area is a two-dimensional measurement, the dimensions are multiplied by a scale that is squared to determine the scale of the corresponding areas. The dimensions of the rectangle are multiplied by a scale of 3. Therefore, the area is multiplied by a scale of 3^2 (which is equal to 9):

$$24cm \times 9 = 216cm$$

Dear CBEST Test Taker,

We would like to start by thanking you for purchasing this practice test book for your CBEST exam. We hope that we exceeded your expectations.

We strive to make our practice questions as similar as possible to what you will encounter on test day. With that being said, if you found something that you feel was not up to your standards, please send us an email and let us know.

We would also like to let you know about other books in our catalog that may interest you.

CSET English

This can be found on Amazon: amazon.com/dp/1628454881

CSET Multiple Subject

amazon.com/dp/1628454504

NES Elementary Education

amazon.com/dp/1628454334

CSET Mathematics

amazon.com/dp/1628454571

We have study guides in a wide variety of fields. If the one you are looking for isn't listed above, then try searching for it on Amazon or send us an email.

Thanks Again and Happy Testing!
Product Development Team
info@studyguideteam.com

Interested in buying more than 10 copies of our product? Contact us about bulk discounts:

bulkorders@studyguideteam.com

FREE Test Taking Tips DVD Offer

To help us better serve you, we have developed a Test Taking Tips DVD that we would like to give you for FREE. **This DVD covers world-class test taking tips that you can use to be even more successful when you are taking your test.**

All that we ask is that you email us your feedback about your study guide. Please let us know what you thought about it – whether that is good, bad or indifferent.

To get your **FREE Test Taking Tips DVD**, email freedvd@studyguideteam.com with "FREE DVD" in the subject line and the following information in the body of the email:

a. The title of your study guide.

b. Your product rating on a scale of 1-5, with 5 being the highest rating.

c. Your feedback about the study guide. What did you think of it?

d. Your full name and shipping address to send your free DVD.

If you have any questions or concerns, please don't hesitate to contact us at freedvd@studyguideteam.com.

Thanks again!

Made in the USA
San Bernardino, CA
09 May 2020

71376039R00082